Capsicum

Capsicum

Breeding Strategies for Anthracnose Resistance

Orarat Mongkolporn

CRC Press is an imprint of the
Taylor & Francis Group, an **informa** business

The book cover photograph is courtesy of Professor Paul W. J. Taylor.

CRC Press
Taylor & Francis Group
6000 Broken Sound Parkway NW, Suite 300
Boca Raton, FL 33487-2742

Printed on acid-free paper

International Standard Book Number-13: 978-1-138-58923-0 (Hardback)

Library of Congress Cataloging-in-Publication Data

Names: Mongkolporn, Orarat, author.
Title: Capsicum : breeding strategies for anthracnose resistance / author: Orarat Mongkolporn.
Description: Boca Raton, FL : CRC Press, Taylor & Francis Group, 2019. | Includes bibliographical references and index.
Identifiers: LCCN 2018010148 | ISBN 9781138589230 (hardback)
Subjects: LCSH: Peppers--Disease and pest resistance. | Peppers--Breeding. | Anthracnose.
Classification: LCC SB608.P5 M66 2019 | DDC 615.3/21--dc23
LC record available at https://lccn.loc.gov/2018010148

Visit the Taylor & Francis Web site at
http://www.taylorandfrancis.com

and the CRC Press Web site at
http://www.crcpress.com

To everyone who initiated, was involved in, and contributed to the chili anthracnose project in the last decade

Contents

Foreword

Capsicum annuum, more commonly known in Southeast Asia and South America as chili and in Australia and Europe as sweet pepper, is a very important global vegetable crop. In Southeast Asia, many farmers grow chili as a fresh vegetable crop or for sending to markets or to factories for processing to make chili sauce. Anthracnose, caused by a complex of *Colletotrichum* species, is the major biotic stress that limits chili production, especially in tropical countries. Anthracnose mainly manifests as a post-harvest disease, resulting in large necrotic lesions on the fruit. This disease is mainly controlled by the application of a "cocktail" of fungicides, as resistant genotypes are not available.

In recent years, insights into the complexity of the pathogen and the genomics of the host have been obtained using cutting-edge molecular technologies. Associate professor Dr. Orarat Mongkolporn has been at the forefront of this technology revolution in *Capsicum* breeding and has actively participated in research to understand both host and pathogen that has led to new genotypes of *Capsicum* containing resistance to anthracnose. No one can be better qualified than Dr. Mongkolporn to summarize and review the literature around this important topic and provide insight into the future of breeding for resistance.

The strength of this book is that it brings together knowledge on both the pathogen and the host, which is often overlooked in similar reviews of an agronomic host or its pathogens. This book will logically inform the reader of the facts behind breeding for resistance from both the host and the pathogen perspective. This information can be used by both researchers and students to enable a constructive debate about *Capsicum* resistance to anthracnose.

The focus of the book addresses an important need for professionals working in *Capsicum* breeding and anthracnose management. They can learn about the complex of species of the pathogen and how to identify each fungal species, and to understand the type of resistance genes involved in the host.

Professor Paul W. J. Taylor
The University of Melbourne

Preface

Developing chili genotypes with resistance to anthracnose has been my life's research work, thanks to Dr. Julapark Chunwongse, who, shortly after I started working at Kasetsart University, provided me with an opportunity to take on this project. I was aware of how economically important the chili crop was to Thailand and the world. I gradually realized how complicated the disease was, which drew my attention to solving the problem. Thanks to Dr. Paul Taylor, professor in plant pathology at the University of Melbourne, for his input and guidance on the pathological aspects of anthracnose.

This lengthy research on improving anthracnose resistance in chili cultivars has built up core knowledge on genetics and breeding for anthracnose resistance and a better understanding of the pathogenicity and host–pathogen interaction. These knowledge and research outcomes have been transformed into this book, which also covers the general aspects of chili as an important global crop, the genomics of chili, and the current molecular technology available to assist plant breeders to develop resistance.

All of the results and outcomes produced from the studies were invaluable resources for writing this book. They would not have been achieved without constant financial support from the Kasetsart University Research and Development Institute, the Thailand Research Fund, and the National Science and Technology Development Agency; hard-working ChiliMarka members; and a strong partnership with the East West Seed Company (Thailand).

Last but not least, I cannot thank my family enough for their unconditional and long-standing support.

Orarat Mongkolporn
Department of Horticulture
Kasetsart University
Kamphaeng Saen Campus
Thailand

chapter one

Introduction

(See color insert.) Colorful and diverse *Capsicum* spp.

1.1 Capsicum *and its global economic importance*

Capsicum, or chili, is a vegetable and spice that has global economic importance. Based on international trade values, chili is ranked as the world's fifth most important vegetable. Two forms, fresh and dry chili, are categorized in global trade by the Food and Agriculture Organization of the United Nations (FAO). The average world chili production from 2012 to 2014 was approximately 31.5 million tons for fresh chili and 3.6 million tons for dry chili (Table 1.1; FAO 2017a). Although chili originated in South America, the world's largest chili producers are in Asia, which contributes 68% of the total chili production in the world (Figure 1.1). China is the world's largest fresh chili producer, with 15.8 million tons per annum—50% of global production. India is the largest dry chili producer, with 1.4 million tons per annum (Table 1.2). Total chili production values in 2014 were US$34 billion, of which 89% was from fresh produce (Figure 1.2).

1.2 *Cultural importance*

Capsicum has several common names, such as pepper, chili pepper, chili, chilli, chile, aji, and paprika, with *pepper* being the most widely used and accepted. This misleading name leads to the assumption that *Capsicum* is "pepper," originally discovered on pepper hunts during the colonial

Table 1.1 Average world fresh and dry chili production during 2012–2014

	Area harvested (ha)		Yield (tons/ha)		Production (tons)	
Region	Fresh	Dry	Fresh	Dry	Fresh	Dry
World	1,941,135	1,875,480	16.2	2.0	31,452,883	3,629,831
Africa	379,554	473,429	8.3	1.7	3,146,817	740,536
Americas	216,695	50,025	18.4	4.7	3,998,856	233,887
Asia	1,233,255	1,287,427	17.3	2.0	21,362,085	2,547,667
Europe	109,266	64,596	26.5	1.7	2,896,543	107,738
Oceania	2,363	2	20.6	1.8	48,582	3.5

Source: Data from FAO, www.fao.org/faostat/en/#data/QC, 2017.

period. However, the actual "pepper" belongs to the genus *Piper* of the Piperaceae family, which is unrelated to the Solanaceae of *Capsicum*. Consequently, in this book, the name *pepper* refers to the genus *Piper* and the name *chili* refers to *Capsicum*.

Chili has a long history that stretches back to the pre-Columbian era. Originating in South America, chili was introduced around the world after being discovered and became a favorite in all areas. Chili is not only economically important but also culturally significant. Chili has transformed the diets of the world and remains embedded in all regional cultures. Chili is the main ingredient in many national cuisines, such as Asian curry dishes, kimchi (Korean pickles), and a variety of chili pastes, sauces, and Mexican and African foods. It is not only limited to savory

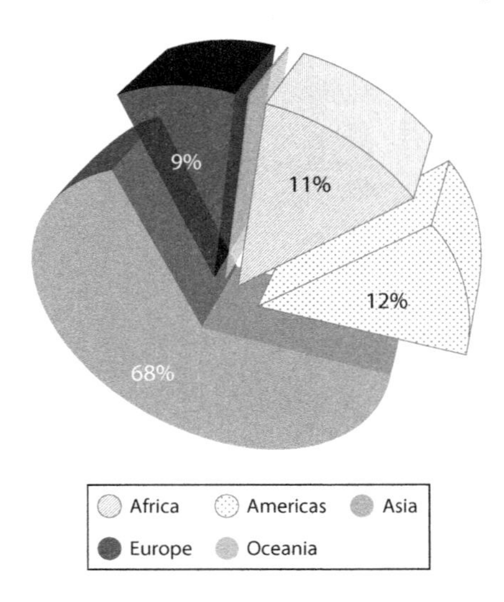

Figure 1.1 Average total fresh and dry chili production by region in 2012–2014. (Data from FAO, www.fao.org/faostat/en/#data/QC, 2017.)

Table 1.2 The world's top 10 producers of fresh and dry chili, average from 2012 to 2014

Country	Fresh chili (tons)	Country	Dry chili (tons)
China	15,840,135	India	1,429,333
Mexico	2,468,924	China	298,957
Turkey	2,109,884	Thailand	259,727
Indonesia	1,752,573	Peru	158,679
Spain	1,039,149	Pakistan	146,727
United States	889,130	Ethiopia	123,589
Nigeria	737,388	Myanmar	118,133
Egypt	605,589	Cote d'Ivoire	116,944
Algeria	480,573	Bangladesh	112,667
Tunisia	377,256	Ghana	103,656

Source: Data from FAO, www.fao.org/faostat/en/#data/QC, 2017.

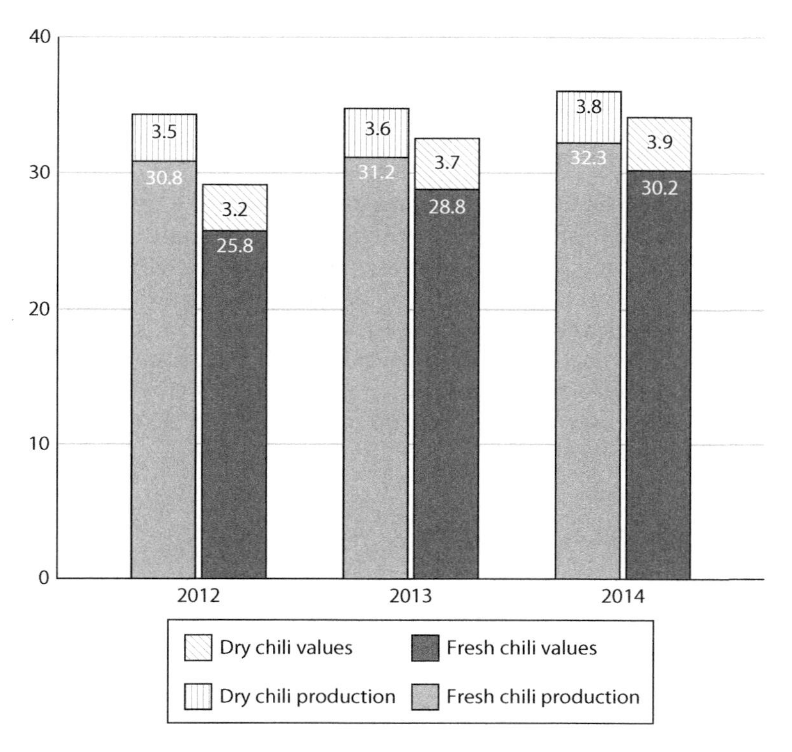

Figure 1.2 Global fresh and dry chili production in 2012–2014. The quantity is expressed in millions of tons, and the value is in US$ billion. (Data from FAO, www.fao.org/faostat/en/#data/QC, 2017; www.fao.org/faostat/en/#data/QV), 2017).

foods; sweets such as chocolate can also have a dash of chili added. Daily consumption of *Capsicum* spices per person in some countries has been reported as 2.5 g in India, 5 g in Thailand, 15 g in Saudi Arabia, and 20 g in Mexico (O'Neill et al. 2012).

1.3 Nutrition and benefits

1.3.1 Vitamins

Nutritionally, chili is rich in vitamins and minerals. The most prominent ones are vitamins A and C as well as folate, which is one of the B vitamins (Kantar et al. 2016). The contents of vitamin A, vitamin C, and folate were measured in around 100 chili cultivars representing various geographic regions of the world, and substantial variations were found. The results revealed that the range was 303–20,840 IU (international units)/100 g for vitamin A, 11.9–196 mg/100 g for vitamin C, and 10–265 μg/100 g for folate. The recommendations for daily consumption by the U.S. Food and Drug Administration are 5000 IU for vitamin A, 60 mg for vitamin C, and 400 μg for folate. Therefore, chili is an invaluable vegetable for the daily human diet.

1.3.2 Carotenoids

Carotenoids are beneficial health compounds that are found in red chili. Carotenoids are lipophilic yellow-orange-red pigments that can also be used as natural food color additives (Arimboor et al. 2015) and textile dyes (Kulkarni et al. 2011). Carotenoids have an antioxidant property that has physiological benefits. The major carotenoid components include β-carotene, α-carotene, and β-cryptoxanthin, which are sources of provitamin A. Carotenoids have been shown to reduce oxidative stress, inhibit cancer cells, and offer protection from cardiovascular diseases, macular degeneration, and cataracts.

1.3.3 Capsaicin and oleoresin

Capsaicin (*trans*-8-methyl-N-vanillyl-6-nonenamide) is a natural alkaloid derived only from plants in the genus *Capsicum* (O'Neill et al. 2012). Capsaicin is a member of the vanilloid compound family, which includes vanillin from vanilla, eugenol from holy basil, bay leaves, and cloves, and zingerone from ginger. Capsaicin is a hydrophobic, colorless, odorless, crystalline compound with the molecular formula $C_{18}H_{27}NO_3$ (Figure 1.3). *Capsicum* extract, which contains a group of capsaicinoids, is called *oleoresin* (Viktorija et al. 2014; Yeung & Tang 2015).

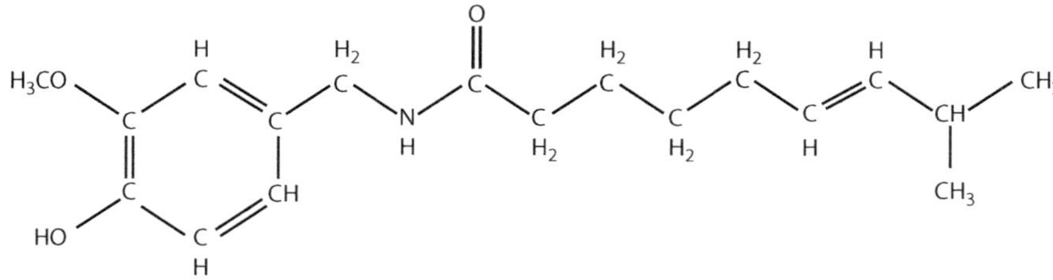

Figure 1.3 Structural formula of capsaicin. (Redrawn from Gudeva, L.K. et al., *Hemijska Industrija* 67, 671–675, 2013.)

Capsaicin is the most commonly occurring of the five major capsaicinoids. Of the total capsaicinoids, capsaicin represents 69%, dihydrocapsaicin 22%, nordihydrocapsaicin 7%, and homocapsaicin and homodihydrocapsaicin 1% (Gudeva et al. 2013). Capsaicin and dihydrocapsaicin are responsible for chili's pungency; they are twice as pungent as nordihydrocapsaicin and homocapsaicin. Capsaicin is an extraordinarily versatile compound with a wide range of applications, from nutrients and pharmaceuticals to bio-pesticides and non-lethal weapons.

- *Capsaicin measurement*: The pungent flavor is a unique characteristic of chili. The pungency level reflects the quantity of capsaicin in the fruit, which was originally quantified in heat units. The chili heat unit is expressed in Scoville heat units (SHU), traditionally measured by a human tasting panel consisting of five people (Bosland & Votava 2012). A chili sample is diluted until the heat is no longer detected, and the amount of dilution is equivalent to the SHU. Current measurement is performed using a scientific instrument, i.e., high performance liquid chromatography (HPLC). However, the SHU is still a popular unit to classify the pungency level of chili (Figure 1.4). The SHU value can be obtained from the HPLC capsaicin content in milligrams per liter or parts per million; 1 mg/L equals 16 SHU.
- *Health benefits*: Capsaicin has been studied extensively for its biological effects, including cardio-protective, anti-lithogenic, and anti-inflammatory effects, analgesia, thermogenic effect, and beneficial effects on the gastrointestinal system, as well as its potential clinical value for pain relief, cancer prevention, and weight loss (O'Neill et al. 2012; McCarty et al. 2015; Srinivasan 2016). The topical application of capsaicin has been proved to alleviate pain resulting from arthritis, postoperative neuralgia, and diabetic neuropathy. Capsaicin inhibits acid secretion, stimulates alkali and mucus secretion, and particularly inhibits gastric mucosal blood flow, thus helping to prevent and heal gastric ulcers. Lactating cows that were fed with capsaicin

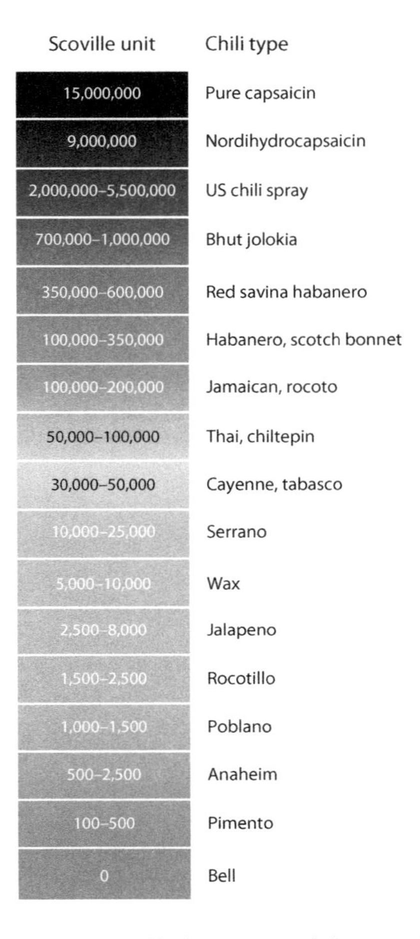

Figure 1.4 (See color insert.) Scoville heat units of diverse chili types. (Redrawn from O'Neill, J. et al., *Pharmacological Reviews*, 64, 939–971, 2012.)

supplements had four times increased milk yield, and their immune system was boosted (Oh et al. 2015).

- *Weapons and bio-pesticides*: Through direct contact with skin, eyes, or mucous membranes (through inhalation or ingestion), capsaicin can cause involuntary closing of the eyes, a burning sensation in the skin, and diminished hand-eye coordination. Therefore, a capsaicin spray has been developed as a non-lethal weapon to control riots and use as a personal self-defense tool (Yeung & Tang 2015). Capsaicin also has antimicrobial properties (Gudeva et al. 2013) and is used in pesticides. Capsaicin is a mammalian irritant, because the substance depolarizes sensory receptor cells and causes pain (Sterner et al. 2005); thus, it can be used to repel rodents.

1.4 Chili types and uses around the world

Chili is an immensely diverse crop. Among the cultivated species, *Capsicum annuum* is the most diverse, with a large variety of fruit shapes, sizes, and pungency levels. More new species continue to be discovered, while some existing ones have been newly classified to become synonyms. However, what happens scientifically does not affect the world of chili trade and consumption. All cultivated and some wild species are used globally for different purposes. The regions where *Capsicum* originated have a large diversity of fruit types that seem to relate to how the chili is used. The following information on common fruit types and their uses (Tables 1.3–1.7) has been gathered from Bosland and Votava (2012), www.cayennediane.com, www.chilipeppermadness.com, and my own experiences.

Table 1.3 (See color insert.) List of *Capsicum annuum*

	Bell: non-pungent, blocky type, 10 cm long. Bell cultivars have ripe fruit in various shades of yellow, orange, red, and brown. Uses: fresh and cooking.
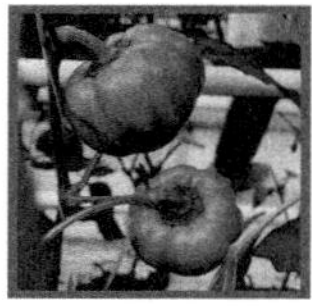	Pimento or pimiento: non-pungent, heart shaped or round, thick wall, red ripe fruit, 7–10 cm long and 5–7 cm wide. Uses: processed foods such as pimento cheese, stuffed olives, or fresh and roasted.
	Wax: non-pungent to mildly pungent, yellow immature fruit turning orange or red when ripe; well known as the long fruit Hungarian wax or banana, 10 cm long and 4 cm wide. Short type is 5×2 cm. Uses: pickles and fresh.
	Ancho: mildly pungent, heart shaped, pointed, thin wall, dark green when immature and red or chocolate brown when ripe, 8–15 cm long; growth is restricted to Mexico. Uses: stuffed foods.
	Pasilla: mildly pungent, long and slender fruit, 15–30 cm long and 2.5–5 cm wide; ripe fruit is brown. Uses: dried to make sauce.

Table 1.3 (Continued)

Cayenne: very pungent, long and wrinkled red ripe fruit, 13–25 cm long and 1.2–2.5 cm wide, widely grown around the world.
Uses: dried, ground, and sauces.

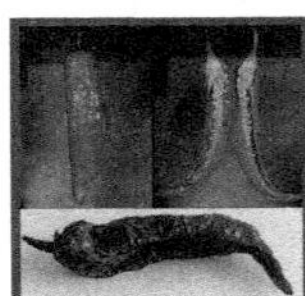

Mirasol: moderately pungent, conical shape, erect fruit has slight curve, 7–10 × 1–2 cm, thin wall, becomes translucent when dried.
Uses: dried.
(Photographs courtesy of Bruno B. Defilippi and Pitchayapa Mahasuk.)

Pepperoncini: non-pungent to mildly pungent, large fruit, 7.5–12.5 cm long, irregular, thin wall.
Uses: pickles.

Jalapeno: very pungent, conical fruit, thick wall, dark green immature fruit, 5–10 cm long and 2.5–3.8 cm wide. Dry fruit skin showing netting pattern or corkiness.
Uses: canned, pickles, and dried.

Serrano: very pungent, cylindrical fruit, 5–10 cm long and 1 cm wide, medium-thick wall, no corkiness.
Uses: salsa sauce.

De Arbol: moderately pungent, long and slender fruit, 5–8 cm long and 0.5–1 cm wide, fruit translucent when dried.
Uses: dried.

Cherry: non-pungent to mildly pungent, small round fruit around 2.5 cm.
Uses: fresh and pickles.

Thai large bird or Khee Noo: very pungent, long fruit, 3–12 cm long.
Uses: fresh, dried, ground, and chili paste.

Table 1.3 (Continued)

	Thai Chee Faa: mildly to moderately pungent, 5–20 cm long fruit, similar to Cayenne but with smooth skin, color yellow to red. *Chee Faa* is Thai for "pointing up to the sky"; however, Chee Faa is not erect. Uses: fresh, dried, ground, chili paste, and sauces.

Sources: Data from Bosland, P., Votava, E.J., *Peppers: Vegetable and Spice* Capsicums, CAB International, Reading, UK, 2012; www.cayennediane.com, 2017; www.chilipepper-madness.com, 2017.

Table 1.4 (See color insert.) List of *Capsicum chinense*

	Habanero: very pungent, lantern shaped, orange or red ripe fruit, 6 cm long and 2.5 cm wide. Uses: fresh, salsa sauce, fermented to make spicy sauce.
	Bhut Jolokia: extremely pungent, lantern shaped, orange or red ripe fruit, 5–7.5 cm long. Uses: defense products.
	Aji dulce: very pungent, fruit size similar to Habanero, red, orange, or yellow fruit, 2.5–5 cm long and 2.5–3.2 cm wide. Uses: fresh, salsa sauce, and fermented to make spicy sauce.
	Charapita: very pungent, very small fruit with 0.6 cm diameter, round and thin flesh, erect, yellow or red ripe fruit. Uses: cooking.
	Biquinho or Chupentinho: moderately pungent, small round fruit around 2.5 cm long with beak-shaped end, bright yellow or red ripe fruit. Uses: ornament, pickles, and cooking.

Sources: Data from Bosland, P., Votava, E.J., *Peppers: Vegetable and Spice* Capsicums, CAB International, Reading, UK, 2012; www.cayennediane.com, 2017; www.chilipepper-madness.com, 2017.

Table 1.5 (See color insert.) List of *Capsicum frutescens*

	Tabasco: very pungent, 2.5–3 cm long fruit, 0.5 cm wide, red ripe fruit. Uses: Tabasco sauce.
	Thai small bird: moderately to very pungent, fruit length <3 cm, erect, white to green immature, red ripe fruit, unique aroma. Uses: fresh, dried, ground, and chili paste. (Photograph courtesy of Kietsuda Luengwilai.)

Source: Data from Bosland, P., Votava, E.J., *Peppers: Vegetable and Spice* Capsicums, CAB International, Reading, UK, 2012.

Table 1.6 (See color insert.) List of *Capsicum baccatum*

	Aji Amarillo: moderately pungent, 10–15 cm long fruit, deep orange when ripe, fruity flavor. Uses: fresh, dried, and paste.

Source: Data from Bosland, P., Votava, E.J., *Peppers: Vegetable and Spice* Capsicums, CAB International, Reading, UK, 2012.

Table 1.7 (See color insert.) List of *Capsicum pubescens*

	Rocoto: very pungent, very thick wall, shape resembles a miniature bell, 5–13 cm long, yellow to red ripe fruit with black seeds. Uses: fresh, dried. and paste. (Photograph courtesy of Bruno B. Defilippi)

Source: Data from Bosland, P., Votava, E.J., *Peppers: Vegetable and Spice* Capsicums, CAB International, Reading, UK, 2012.

Capsicum remains one of the most important crops in the world, with uses as a vegetable, a spice, and a medicine, and in bio-weapons and bio-pesticides. This high demand for *Capsicum* products attracts scientists globally to continue researching this crop for human needs. This book focuses on chili anthracnose, which has long been a major problem of chili production in Asia, the world's largest chili producer. There has been little success in improving chili resistance to anthracnose. This book will incorporate core research findings on anthracnose resistance covering genetics, molecular breeding, and pathology.

chapter two

Capsicum *genome, origin and diversity*

2.1 Capsicum *genomes—current status*

2.1.1 Genome sequences

Capsicum belongs to the Solanaceae family, is diploid (*n* = 12 and 13), and is a facultative self-pollinating crop. The *Capsicum* genome is complex and contains a large amount of repetitive DNA, which causes inflation in the genome size. Initially, before its genome sequencing, the *Capsicum* genome size was estimated by flow cytometry to range from 3.09 to 5.64 Gb (10^9 base pairs; Park & Choi 2013). The smallest *Capsicum* genome belongs to *C. annuum*, while the largest is *C. parvifolium* (Moscone et al. 2003). To date, five *Capsicum* genomes have been completely sequenced, of which three were *C. annuum* and two were *C. baccatum* and *C. chinense* (Kim et al. 2014; Qin et al. 2014; Kim et al. 2017).

C. annuum cv. "CM334," a Mexican landrace, is one of the *Capsicum* genome references. "CM334" is highly resistant to a diverse range of pathogens including *Phytophthora*, *Capsicum* mottle virus, and nematodes (Wang & Bosland 2006). The genome size of "CM334" is 3.06 Gb with 34,903 genes (Figure 2.1; Kim et al. 2014). Whole genomes of a different cultivar of *C. annuum*, "Zunla-1," and a wild *C. annuum* var. *glabriusculum* accession have also been completely sequenced, revealing their larger genome sizes of 3.35 and 3.48 Gb, respectively (Qin et al. 2014). A more recent study revealed that *C. chinense* and *C. baccatum* genomes were 3.2 and 3.9 Gb, respectively (Kim et al. 2017). The increased size of *C. baccatum* genome contains approximately 35,000 genes, similarly to the other smaller *Capsicum* genomes as displayed in Table 2.1.

Interestingly, *Capsicum* genomes contain very large numbers of transposable elements, greater than 75% of the whole genome, most of which are retrotransposons, particularly long terminal repeat (LTR) retrotransposons. General features of the three *C. annuum* reference genomes are displayed in Table 2.1. *Capsicum* is considered to have a large-sized genome that does not experience whole-genome duplication. The genome expansion is mainly due to DNA repeats, comprising both the LTR retrotransposons and long unit tandem repeats (Park et al. 2012).

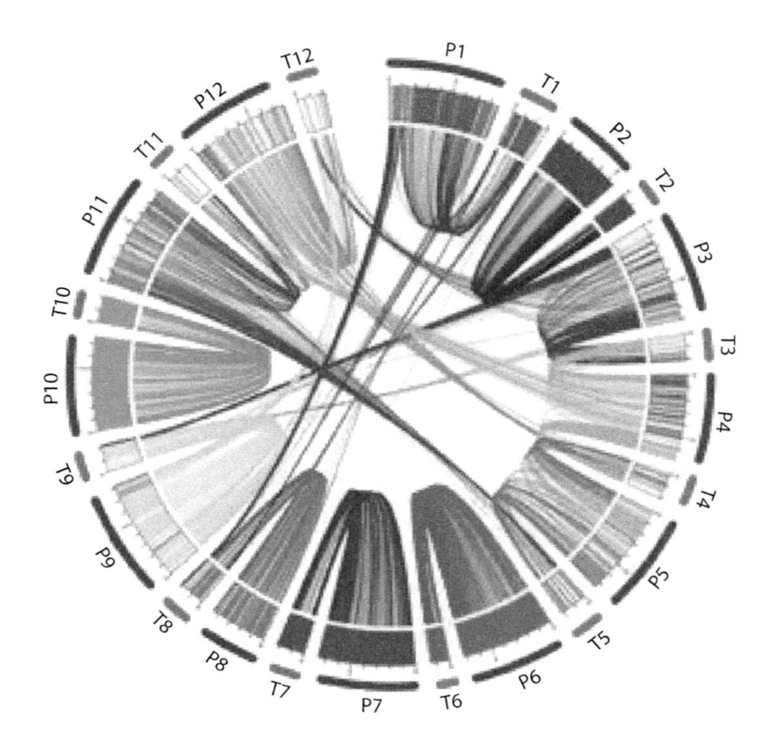

Figure 2.1 (See color insert.) *Capsicum* genome, *C. annuum* "CM334," compared with the tomato genome. P1–P12 represent *Capsicum* chromosomes, and T1–T12 represent tomato chromosomes. Lines link the orthologous genes between the two genomes. (Modified from Kim, S. et al., *Nat. Genet.*, 46, 270–278, 2014. With permission.)

Table 2.1 General features of three *Capsicum annuum* genomes in comparison with the tomato genome

Features	CM334[a]	Zunla-1[b]	Chiltepin[b]	Tomato[a,b]
Genome size (Gb)	3.06	3.48	3.35	0.76
GC content (%)	35.03	34.9	35.0	34.0
Number of genes	34,903	35,336	34,476	34,771
Average gene length (bp)	1,159	3,363	3,235	3,006
Average exon no./gene	1.14	4.27	4.04	4.6
LTR/TE (%)	72.6	70.3	70.1	50.3
Total TE (%)	76.4	80.9	81.4	61.3
Chromosome number (*n*)	12	12	12	12

LTR, LTR retrotransposon; TE, transposable element.
CM334 is a Mexican landrace; Zunla-1 is a *C. annuum* cultivar; Chiltepin is a wild *C. annuum* var. *glabriusculum* accession.
[a] Kim et al. (2014)
[b] Qin et al. (2014)

LTR retrotransposons are a major evolutionary force in animals, fungi, and plants that causes genomic instability. LTR retrotransposons facilitate the creation of new candidate genes called *retrogenes* through retroduplication, in which spliced mRNA is captured, reverse transcribed, and subsequently integrated into the genome by a retrotransposon. Unique characteristics of a retrogene include a reduced number of introns compared with its parental gene sequence and the presence of a 3′ poly (A) tail and flanking direct repeats (Kim et al. 2017).

2.1.2 Genetic and physical maps

Genetic and physical maps of *C. annuum* were compared by Han et al. (2016b). An ultra-high-density SNP map was constructed from recombinant inbred lines of *C. annuum* "Perennial" and "Dempsey" and compared with the reference physical map of "CM334" (Figure 2.2; Table 2.2). Inconsistencies of bin order were detected in all chromosomes. Extensive inconsistencies were found on the upper region of P8. A bin is a position on a genetic map with a unique segregation pattern that is separated from adjacent bins by a single recombination event. In this particular map, all SNPs were grouped into 2578 bins. The average length of bins was 1.07 Mb (100 kb to 141 Mb).

Chili and tomato are members of the Solanaceae family, sharing the same ancestor before their divergence around 19.6 million years ago (Wu & Tanksley 2010). The *Capsicum* genome is four times larger than the tomato genome. The gene repertoire of the two species was highly conserved, but the linear order of the genes was greatly modified, and the recombination rates of the two genomes were not significantly different (Tanksley et al. 1988). A synteny map was constructed with COSII markers using an interspecific population of *C. frutescens* "BG2814-6" and *C. annuum* cv. "NuMex RNaky" (Wu et al. 2009). COSII markers are conserved orthologous single copy genes identified in the Solanaceae and Rubiaceae and have been widely used in comparative mapping studies among the key solanaceous species (Wu and Tanksley 2010). The markers used in the map were orthologous between *Capsicum* and tomato. The 12 *Capsicum* chromosomes are designated P1–P12 based on the synteny with tomato chromosomes T1–T12. *Capsicum* and tomato comparative maps identified 19 inversions and six translocations in most chromosomes except for P7 and P8. The synteny map characterized a reciprocal translocation that differentiated the cultivated *C. annuum* genome from the wild *C. annuum* and other *Capsicum* species genomes. The chromosome translocation involved crossing over between two non-homologous chromosomes of wild *C. annuum*, *C. frutescens*, and *C. chinense*, referred to as *P1-wild* and *P8-wild*. As a result, P1 and P8 of two recent maps were represented by four chromosomes designated P1-wild, P8-wild, P1-cultivated, and P8-cultivated (Hill et al. 2015; Figure 2.3). Both maps were based on ultra-high-density

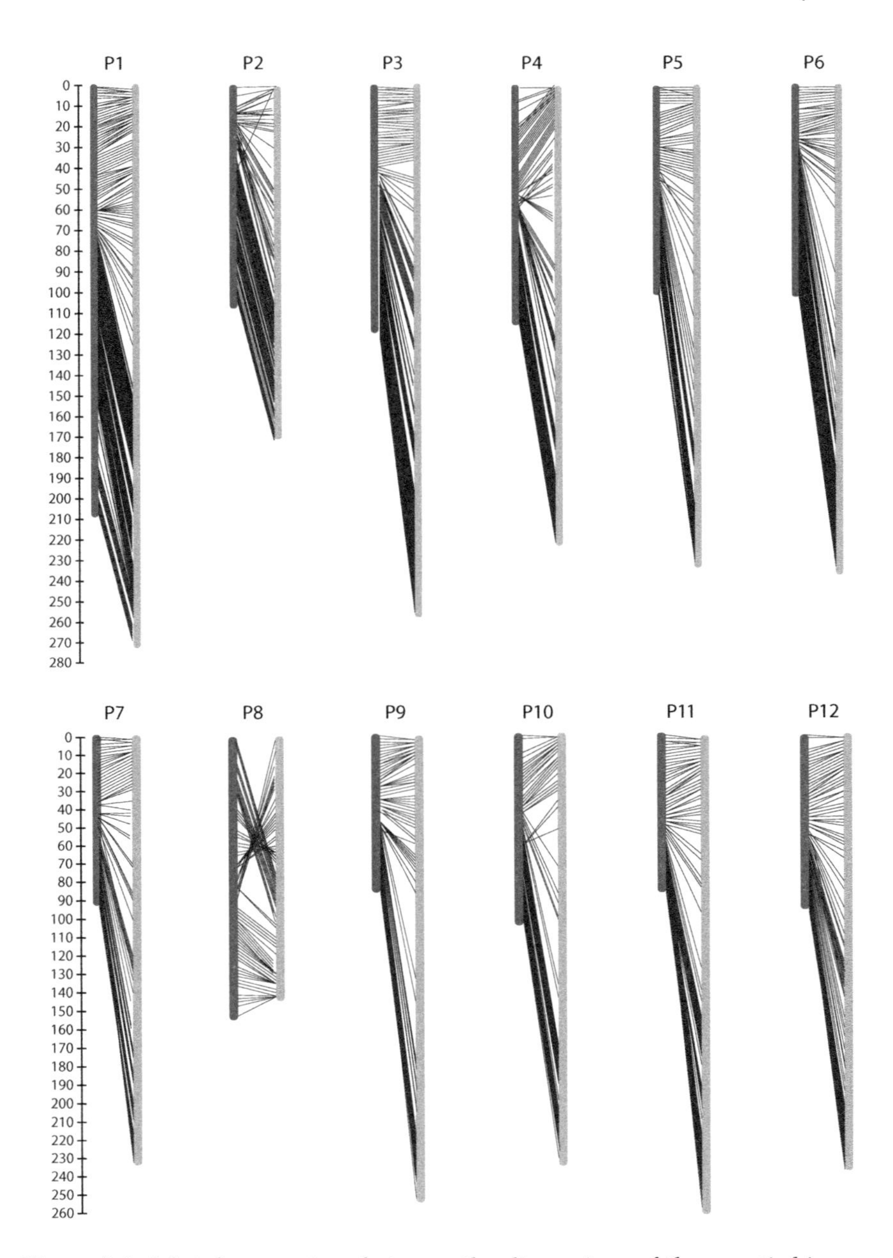

Figure 2.2 A brief comparison between the dimensions of the genetic bin map of *Capsicum annuum* (left chromosomes) and the reference physical map of *C. annuum* "CM334" (right chromosomes) shows inconsistencies of bin order in all chromosomes. Lines link common SNP markers on the two maps. Scales are map distance in either centimorgans for genetic or megabases for physical maps. (Manually redrawn from Han, K. et al., *DNA Res.*, 23, 81–91, 2016.)

Table 2.2 Genetic and physical distances in the bin map

Chromosomes	SNP number	Bin number	Genetic map (cM)	Physical map (Mb)
P1	82,996	370	208.5	272.6
P2	80,141	195	107.5	171.1
P3	87,793	261	118.5	257.9
P4	54,657	216	116.5	222.5
P5	82,413	190	100.6	233.4
P6	107,015	220	102.6	236.9
P7	84,339	175	92.5	231.9
P8	24,383	217	153.7	144.8
P9	275,842	161	86.2	252.7
P10	230,360	154	103.9	233.6
P11	252,756	196	86.8	259.7
P12	68,549	223	94.9	235.7
Total	1,431,244	2,578	1,372.2	2,752.8

Source: Modified from Han, K. et al., *DNA Res.*, 23, 81–91, 2016.

expressed sequence tags (ESTs), which were constructed from recombinant inbred line (RIL) populations. One was developed from the same cross "BG2814-6" and "NuMex RNaky," and the other was intraspecific *C. annuum* "Early Jalapeno" and "CM334." RILs are homozygous F6:F7 or F7:F8 derived from single seed descent of F2 lines.

A more recent study, based on the genome structure comparison of three *Capsicum* genomes, *C. annuum*, *C. baccatum*, and *C. chinense*, has revealed significant chromosomal translocations among chromosomes 3, 5, and 9 that differentiated *C. baccatum* from the other two *Capsicum* species (Figure 2.4). Significant translocations were detected between the distal region on the long arm of *C. baccatum* P9 and the short arm of P3 in a *C. annuum*/*C. chinense* common ancestor, and between the terminal region of the short arm of *C. baccatum* P3 and the long arm of *C. annuum*/*C. chinense* P9.

2.1.3 *Synteny study*

The chromosomal evolution in the Solanaceous species, including tomato, potato, eggplant, chili, and tobacco, was studied using comparative COSII maps (Wu & Tanksley 2010). The most recent common ancestors were estimated by phylogenetic relationships with molecular dating analyses. The most common ancestors of tobacco and tomato possibly lived 23.7 million years ago; of chili and tomato, 19.6 million years ago; of eggplant and tomato, 15.5 million years ago; and of potato and tomato, 7.3 million years ago. The tomato genetic map was used as a reference due

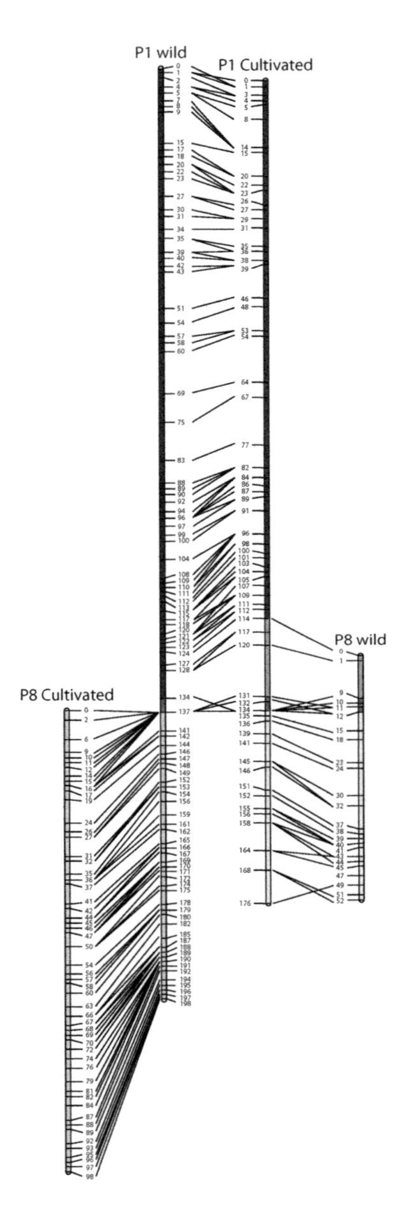

Figure 2.3 (See color insert.) *Capsicum* chromosomes show recombination on P1 and P8. Chromosomal regions are colored according to translocation pairs. Non-translocated portions of P1 (P1 wild/P1 cultivated) are in blue, the translocated arms P8 wild/P1 cultivated are in aqua, and P8 wild/P8 cultivated are in yellow. Grey corresponds to pseudolinkage between P1 and P8 wild; red indicates the translocation break point. (Modified from Hill, T. et al., *G3*, 5, 2341–2355, 2015. Figure licensed under the Creative Commons Attribution 4.0.)

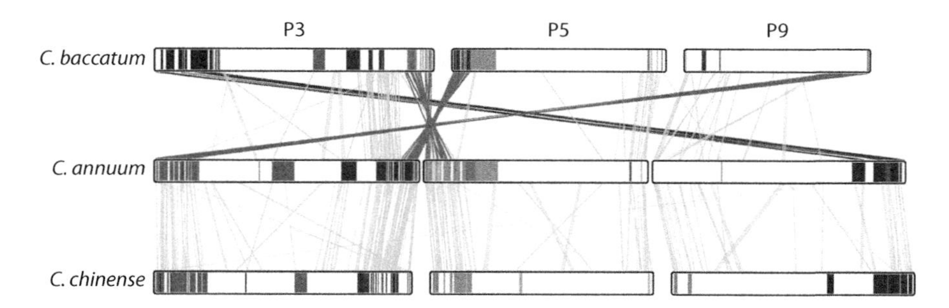

Figure 2.4 (See color insert.) A linear comparison of chromosomal rearrangements in three *Capsicum* genomes. Line colors indicate translocations in the ancestral lineage of *C. annuum* and *C. chinense* (red), in *C. baccatum* (green), and in ancestors of *C. annuum* and *C. chinense* or *C. baccatum* (dark grey). (Modified from Kim, S. et al., *Genome Biol.*, 18, 210, 2017. Figure licensed under the Creative Commons Attribution 4.0.)

to its highly dense markers. Among three cultivated *Capsicum* species, *C. annuum*, *C. baccatum*, and *C. chinense*, whose genomes have been completely sequenced (Kim et al. 2017), phylogenetic analysis suggested that *C baccatum* diverged from the common progenitor of *C. annuum* and *C. chinense* 1.7 million years ago, and the divergence between the latter two species occurred later, around 1.1 million years ago.

Across the family, rates of chromosomal rearrangement were estimated to be from 0.1 to 1 inversions and 0.2 to 0.4 translocations per million years. The Solanaceae experienced chromosomal changes at a modest rate compared with other families. Inversions appeared to occur at a higher rate than translocations per million years.

Figure 2.5 exhibits chromosomal rearrangements of solanaceous crop species. In summary, since all the species diverged from their most recent common ancestor, four inversions have occurred in tomato, two in potato, 16 in eggplant, and one in chili, although there were 11 undetermined inversions. Regarding the translocations, a T5 (tomato chromosome 5) segment inserted into E3 (eggplant chromosome 3), a T10 segment into E10, and T10 into E4 and E10. For *Capsicum* translocations, a non-reciprocal translocation resulted in P1 and P8, a small T4 segment inserted into P3. Undetermined translocations were in chromosomes 4, 5, 11, and 12 for potato and eggplant, and in chromosomes 3, 4, 5, 9, 11, and 12 for eggplant and chili.

2.1.4 *Chloroplast and mitochondria genomes*

Complete chloroplast genome sequences have been achieved from four cultivated *Capsicum* species: two varieties of *C. annuum*, including *C. annuum* var. *annuum* and var. *glabriusculum* (Raveendar et al. 2015a,b); *C. frutescens* (Shim et al. 2016); *C. chinense* (Park et al. 2016); and *C. baccatum* var. *baccatum*

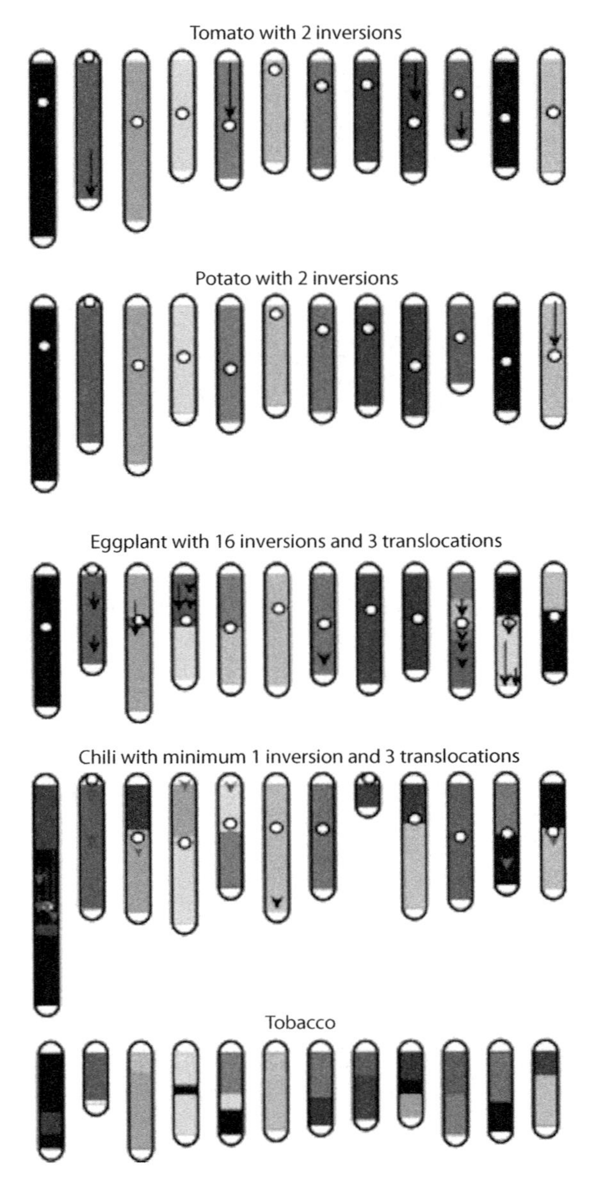

Figure 2.5 (See color insert.) Chromosomal evolution in the Solanaceae. Each tomato chromosome is assigned a different color that matches the orthologous counterparts in other species, thus depicting the translocations differentiating these species. Black arrows on the chromosomes indicate inversions; grey arrows represent uncertain inversions. White dots represent putative centromere locations. (Modified from Wu, F., Tanksley, S.D., *BMC Genomics*, 11, 182, 2010. Figure licensed under the Creative Commons Attribution 2.0.)

(Kim et al. 2016). The genome sizes of chloroplast from *C. annuum*, *C. frutescens*, and *C. chinense* are similar, with approximately 156.8 kb and a maximum 61 bp difference. The *C. baccatum* chloroplast genome appears to be the largest (157 kb). All genomes have a similar GC content (37.7%). The common chloroplast genome structure comprises a pair of inverted repeats (IRs) that are separated by a small and a large single copy (SSC and LSC) regions and simple sequence repeats (SSRs) (Figure 2.6; Table 2.3). All chloroplasts harbor 132 known genes, except for those of *C. chinense*, which harbor 113 genes. Phylogenetic analysis reveals that *C. chinense* is the most closely related to *C. annuum* var. *glabriusculum*, which is a wild progenitor of *C. annuum* (Park et al. 2016). Therefore, the chloroplast genome

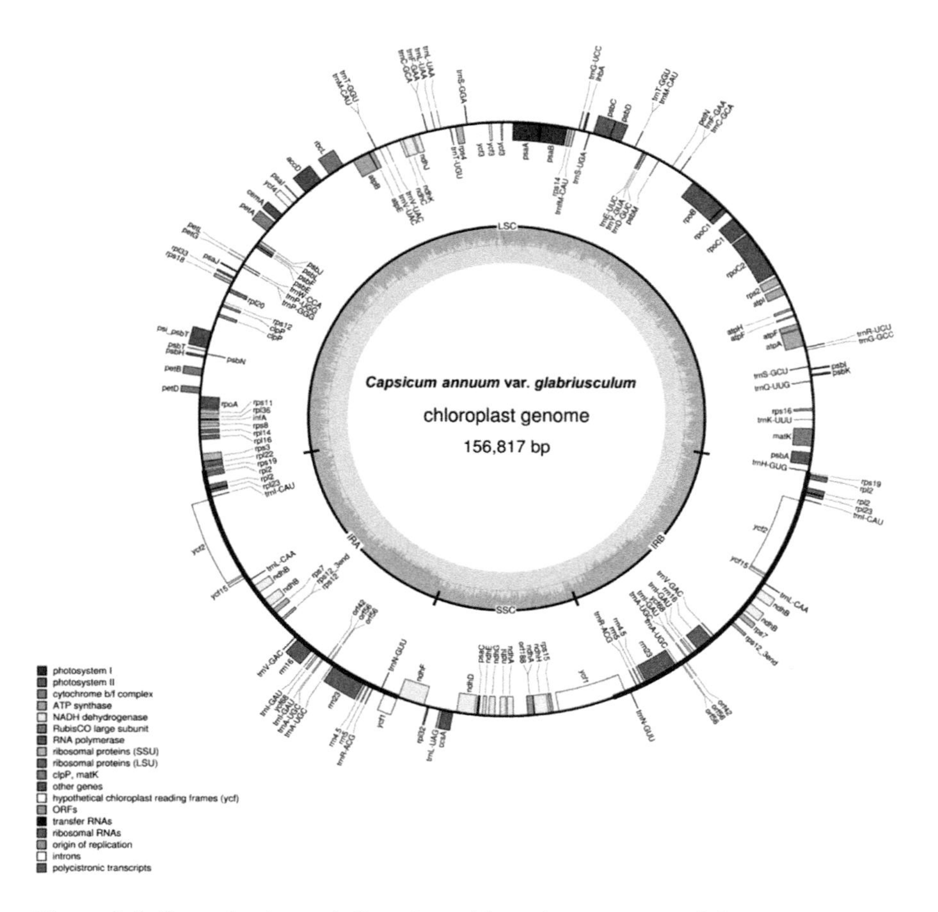

Figure 2.6 (See color insert.) Complete chloroplast genome of *Capsicum annuum* var. *glabriusculum*. Inner genes are transcribed clockwise, and outer genes are transcribed counterclockwise. (From Raveendar, S. et al., *Molecules*, 20, 13080–13088, 2015. Figure licensed under the Creative Commons Attribution 4.0.)

Table 2.3 General features of *Capsicum* chloroplast genomes

		Size (Base Pairs)				Number of Genes		
Capsicum **Species**	GC content (%)	Genome	IRs	SSC	LSC	Total	Duplicate in IRs	SSR Motifs
C. annuum var. *annuum*[a]	37.7	156,878	51,602	17,929	87,347	132	7	144
C. annuum var. *glabriusculum*[b]	37.7	156,817	50,284	18,948	87,446	132	7	125
C. frutescens[c]	37.7	156,817	51,584	17,853	87,380	132	7	125
C. chinense[d]	37.7	156,807	51,606	17,911	87,290	113	28	na
C. baccatum var. *baccatum*[e]	37.7	157,145	51,820	17,974	87,351	132	7	134

IRs, inverted repeats; LSC, large single copy region; na, not available; SSC, small single copy region; SSR, simple sequence repeat.

[a] Raveendar et al. (2015a).
[b] Raveendar et al. (2015b).
[c] Shim et al. (2016).
[d] Park et al. (2016).
[e] Kim et al. (2016).

can effectively identify species and can be used for phylogenetic studies. Table 2.3 displays general features of the five chloroplast genomes.

The mitochondrial genome has been completely sequenced in *C. annuum* to study the cytoplasmic male sterility (CMS) trait (Jo et al. 2014). Structural variations in mitochondrial DNA are associated with several mutant phenotypes such as CMS, which is an economically important trait for the hybrid seed industry. Mitochondrial genomes of two chili genotypes, CMS, and restorer lines were compared (Table 2.4). The CMS "FS4401" and fertile "Jeju" mitochondrial DNAs contained a similar complement of protein coding genes except for one gene, *atp6*, in "FS4401." Extensive DNA rearrangements were detected in each genome, which were identified in 16 syntenic blocks on each genome (Figure 2.7). The sizes of the blocks ranged from 2.9 (Block 10) to 78.9 kb (Block 15). Block 6 and a part of Block 1 were duplicated in FS4401, and Block 7 and parts of Blocks 4 and 6 were duplicated in "Jeju." The sequences between blocks were unique to each mitochondrial genome. Previously reported CMS candidate genes were located on the edges of the highly rearranged CMS-specific DNA regions and near the DNA repeats. Interestingly, these characteristics were also found in the CMS-associated genes of other species, such as maize, sugar beet, and radish, suggesting that a common mechanism could have been involved in the evolution of the CMS-associated genes.

2.2 Capsicum *origin and distribution, and species identification*

Capsicum is a member of the Solanaceae family (tribe Solaneae, subtribe Capsicinae), which also includes tomato, potato, tobacco, and petunia. *Capsicum* is native to the tropics and subtropics of America from

Table 2.4 General features of *Capsicum* mitochondrial genomes in comparison to tobacco

Features	FS4401	Jeju	Tobacco
Genome size (bp)	507,452	511,530	430,597
GC content (%)	44.5	44.6	45.0
Coding sequences (bp)	40,085 (7.9%)	39,524 (7.7%)	43,642 (10.1%)
Repeated sequences (bp)	42,505 (8.4%)	70,122 (13.7%)	73,511 (17.1%)
Gene number	66	64	61
Protein coding genes	38	37	37
rRNAs	3	3	3
tRNAs	12	12	12

Source: Modified from Jo, Y.D. et al., *BMC Genomics*, 15, 561, 2014.

FS4401 is a CMS *Capsicum annuum* line; Jeju is a fertile *C. annuum* line. tRNA, transfer RNA.

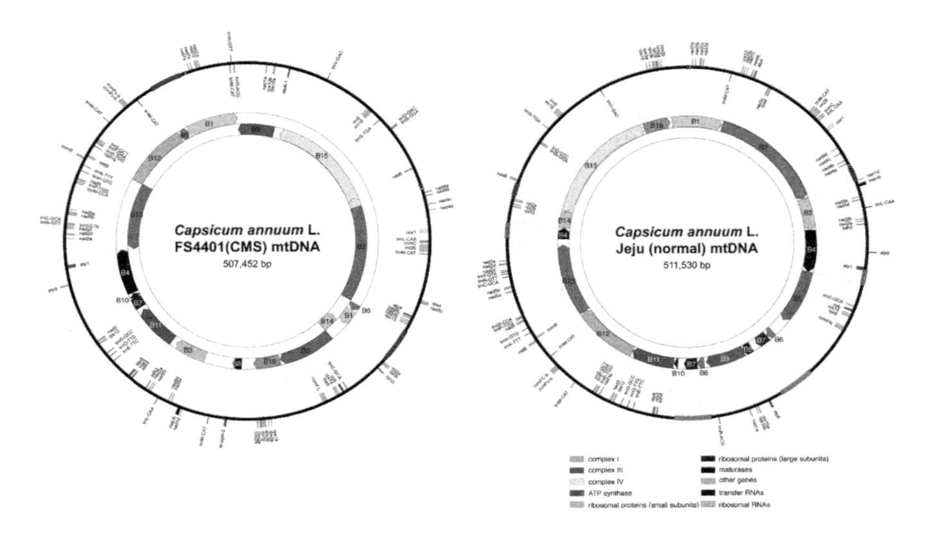

Figure 2.7 (See color insert.) Maps of mitochondrial genomes of *Capsicum annuum*, CMS "FS4401" and fertile "Jeju" lines, displaying 16 syntenic sequence blocks (inner circles), B1–B16, between genomes (>95% similarity). Two inner circles are depicted in the CMS "FS4401" to separate blocks with different directions. Different colors denote different functions of the gene products. (Modified from Jo, Y.D. et al., *BMC Genomics*, 15, 561, 2014. Figure licensed under the Creative Commons Attribution 4.0.)

pre-Columbian times, referred to as the "New World" in literature. *Capsicum* centers of distribution are thought to be from the southern United States and Mexico to western South America, northeastern Brazil, and coastal Venezuela, eastern coastal Brazil, and from central Bolivia and Paraguay to northern and central Argentina (Hunziker 2001). Based on archaeological evidence, *Capsicum* was domesticated at least 6000 BP (before present) (Perry et al. 2007). Among the domesticated *Capsicum* species, *C. annuum* is the most important and is widely grown around the world.

According to a review of *Capsicum* taxonomy by Eshbaugh (2012), the genus was previously described as possessing campanulate corollas and the presence of capsaicin. However, two wild species, *C. lanceolatum* and *C. rhomboideum*, were notably lacking pungency (Moscone et al. 2007). Five domesticated, or cultivated, species were classified based on morphology of corolla, calyx, and seed in the twentieth century, including *C. annuum* var. *annuum*, *C. chinense*, *C. frutescens*, *C. baccatum* var. *pendulum*, and *C. pubescens*. *C. pubescens* is the most distinct due to its black seed and hairy leaves; thus, it is easier to recognize than the other four species. *C. annuum*, *C. chinense*, and *C. frutescens* form a complex that evolved in the tropical lowlands of Latin America and the Caribbean (Eshbaugh 2012), with *C. annuum* the dominant species in Mexico, *C. chinense* in Amazonas, and *C. frutescens* in the Caribbean (Figure 2.8). The three species are more widely grown

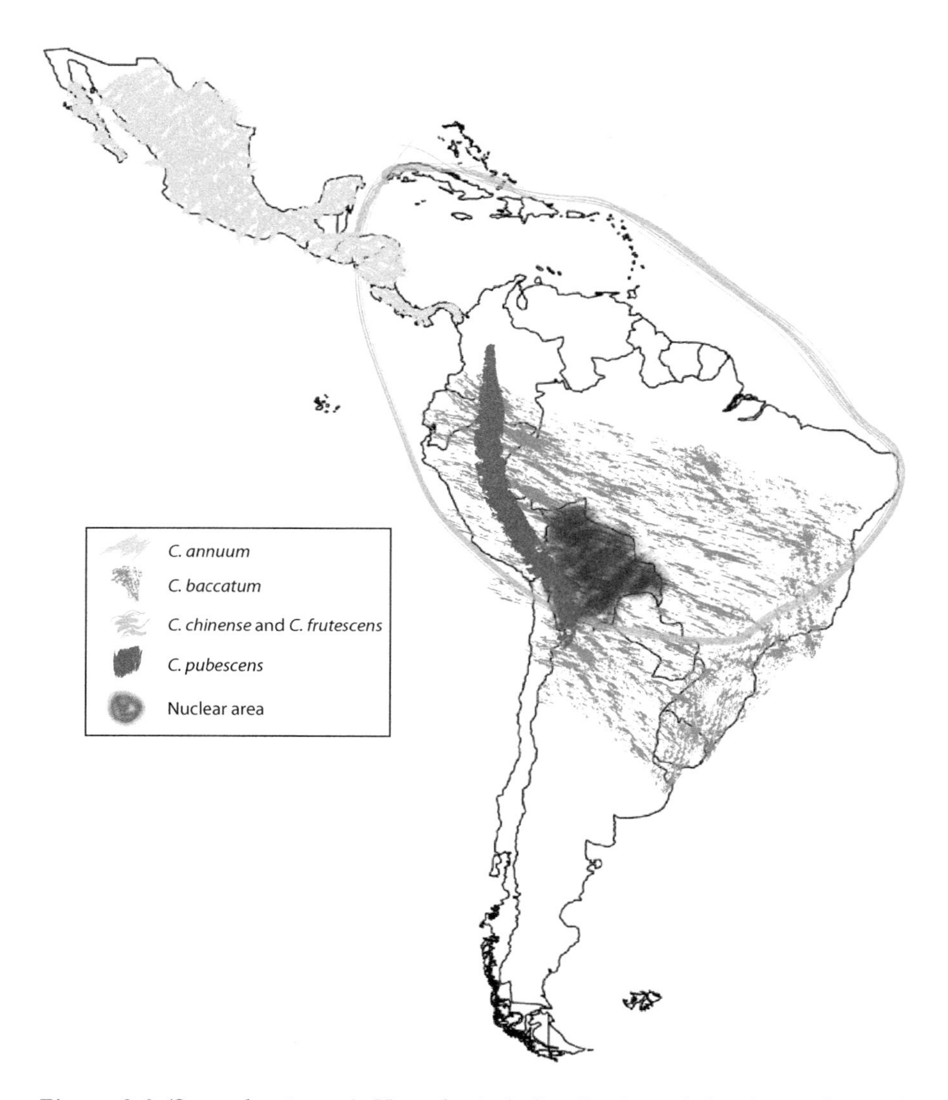

Figure 2.8 (See color insert.) Hypothetical distribution of the five cultivated *Capsicum* during the discovery of the New World. (Redrawn from Eshbaugh, W.H., *Peppers: Botany, Production and Uses*, CAB International, Wallingford, UK, 2012.)

than *C. baccatum* and *C. pubescens*, which are limited to South America. *C. annuum* was introduced to Europe by Columbus and subsequent explorers of Mesoamerica, while the Portuguese explorers were responsible for the introduction of *C. chinense* to Eastern Europe, Africa, and Asia.

There has been a debate as to the speciation of the three closely related *Capsicum* species, *C. annuum*, *C. frutescens*, and *C. chinense*. *C. frutescens* was excluded from the Plant List, version 1.1 (2013), because *C. frutescens* was claimed to be a synonym of *C. annuum*. In contrast, a current study on genetic diversity of Thai chili landraces using genome-wide single nucleotide polymorphism (SNP) analysis by W. Kethom and O. Mongkolporn (unpublished) has revealed that the two species *C. annuum* and *C. frutescens* are highly distinct (Figure 2.9), which supports the idea that the two species are different.

C. frutescens and *C. chinense* are not easily differentiated based on morphological characters; thus, they may be considered to be the same species. However, evidence supports species distinction based on morphological differences, molecular phylogenetics, and hybrid fertility (Baral & Bosland 2004). Morphological characters, including calyx constriction and flower position, differentiate *C. frutescens* and *C. chinense*. Molecular analysis separated all the chili accessions according to species. Their hybrid progenies showed reduced fertility. Hunziker (2001) recognized only three domesticated species: *C. annuum*, *C. baccatum*, and *C. pubescens*. Pickersgill (1971) found that differences between *C. chinense* and *C. frutescens* are much smaller than those between other domesticated–wild species pairs,

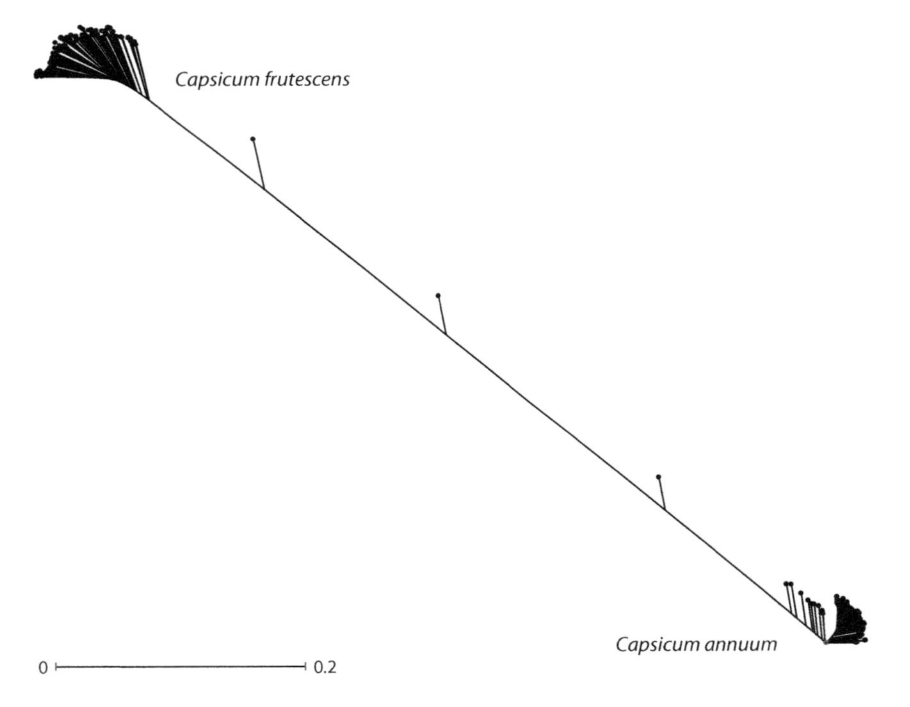

Figure 2.9 A dendrogram of 250 Thai chili landraces generated by *Capsicum* genome-wide SNPs; two distinct groups of *C. annuum* and *C. frutescens* are formed. (Dendrogram courtesy of Wassana Kethom.)

such as *C. annuum* var. *annuum* and *C. annuum* var. *glabriusculum*, or *C. baccatum* var. *pendulum* and *C. baccatum* var. *baccatum*, or *C. pubescens* and *C. eximium*. Nevertheless, five distinct domesticated species continue to be recognized globally by horticultural researchers and commercial traders (Eshbaugh 2012).

C. annuum var. *annuum* L. has a variety of fruit sizes, from small fruit such as "bird chili" to large fruit such as "bell chili" (see Chapter 1). Well-known chili types in this group are jalapino, pablano, Anaheim, ancho, bell, cayenne, and serrano. *C. annuum* var. *glabriusculum* was proposed to be the wild ancestor of this species (Heiser & Pickersgill 1975).

C. chinense is the most economically important chili grown in the Caribbean. It is popular for its pungency and is widely used to add taste and flavor to Caribbean cuisine. *C. chinense* does not have any true wild form. Approximately 200 accessions of *C. chinense* have been identified, which represent different geographical regions of South America comprising the two major zones of Latin America and the Caribbean (Moses & Umaharan 2012). Molecular analysis based on random amplified polymorphic DNA (RAPD) markers grouped chili genotypes into three clusters according to region. The chili from the Caribbean and Lower Amazon of Latin America formed separate groups, and the rest formed the largest group; these were from both Latin America (Upper Amazon and Central America) and the Caribbean (Lesser Antilles, Trinidad, and Tobago). The greatest diversity of chili genotypes was from the Upper Amazon.

C. baccatum is commonly known as *aji* or Peruvian hot pepper. Its center of origin is believed to be near Bolivia and southern Peru, but molecular clustering suggests Paraguay (Albrecht et al. 2011). *C. baccatum* comprises both wild and cultivated forms, var. *baccatum* and var. *pendulum*, respectively. The wild species typically bears small, erect, and deciduous fruit, while the cultivated species typically bears large, pendant, and persistent fruit. Wild *C. baccatum* var. *baccatum* is the progenitor of the domesticated *C. baccatum*. *C. baccatum* has a wide geographic distribution across South America, from the west to the east coast and from north to south, which contains a wide range of climates, from cool Andean highland to Amazonian rain forest. Therefore, with this broad range of geography and climate, *C. baccatum* has greater genetic diversity (Albrecht et al. 2012). Molecular analysis by amplified fragment length polymorphism (AFLP) revealed that the genetic diversity of the wild *C. baccatum* was greater than that of the cultivated varieties (Albrecht et al. 2011). The cultivated *C. baccatum* was divided into two groups, one of which was predominantly in the east, i.e., Paraguay and eastern Argentina, while the other was distributed throughout Peru, Colombia, Ecuador, Bolivia, Chile, and northwestern Argentina. Figures 2.10 and 2.11 display the diversity of the cultivated *Capsicum* and some wild species.

Figure 2.10 (See color insert.) Diversity of *Capsicum annuum* complex: flowers and fruits. *C. annuum* var. *annuum* (1–4); *C. annuum* var. *glabriusculum* (5–8); *C. chinense* (9–13), showing the famous Habanero (11) and Bhut Jolokia (12), typical large leaves of the species (13); and *C. frutescens* (14–16) showing typical exerted stigma (14).

Figure 2.11 (See color insert.) Diversity: *C. baccatum*, *C. pubescens*, and some wild species. *C. baccatum* var. *pendulum* (1–3), *C. baccatum* var. *baccatum* (4,5), *C. baccatum* var. *praetermissum* (6), *C. pubescens* (7,8), *C. chacoense* (9), *C. eximium* (10), and *C. flexuosum* (11,12).

The flower characteristics of domesticated species are described in Table 2.5. A modified taxonomic key to identify domesticated *Capsicum*, developed from IBPGR (1983) by Eshbaugh (2012), is as follows:

1. Corolla violet with white center (see also *C. annuum*); seed dark brown to black, prominently reticulate; anthers purple to violet; calyx with 5 conspicuous deltoid teeth about 1 mm long .... ***C. pubescens***
2. Corolla various shades of white to yellow, with or without distinct markings; seed cream to yellow; anther blue to purple, rarely yellow; calyx teeth present or absent **2**
   3. Calyx bearing distinct elongate teeth usually longer than 0.5 mm; corolla off white (cream) with a pair of yellowish, greenish, or tan markings at the base of corolla lobe; calyx with distinct teeth 0.5–1.5 mm long; anthers yellow .... ***C. baccatum***
   2. Calyx teeth lacking, or if present rarely longer than 0.5 mm; anthers blue to purple **3**
      3. Flowers usually 1/node after the first flowering node, rarely more; prominent constriction lacking between base of calyx and pedicel; corolla pure white to bluish white or rarely violet .... ***C. annuum***
      3. Flowers 2 or more/node above the first flowering node (look for scars), very rarely fewer; prominent constriction between base of calyx and pedicel present in both flowering and fruiting or absent **4**
         4. Constriction between base of calyx and pedicel present; corolla dull white; fruit usually pendant. Persistent, firm-fleshed; style exerted not more than 1 mm beyond the anthers .... ***C. chinense***
         4. Constriction between base of calyx and pedicel absent; corolla greenish white; fruit usually erect, deciduous, soft-fleshed; style exerted 1.5 mm or more beyond the anthers .... ***C. frutescens***

To date, 40 *Capsicum* species have been accepted by the Plant List, version 1.1 (2013). The list of the 40 *Capsicum* spp., with their distribution and chromosome number, is provided in Table 2.6. *C. eshbaughii* was renamed from *C. eximium* var. *tomentosum* (Barboza 2011). Two new endemic species have been identified in northeastern Brazil: *C. caatingae* and *C. longidentatum* (Barboza et al. 2011). Their chromosome number (2*n*) is 24. However, these two species have not yet been accepted by the Plant List. All wild chilies have small red berry-like fruit that is attractive to birds (Walsh & Hoot 2001). The following taxonomic key to identify wild *Capsicum* species was developed by Barboza and Bianchetti (2005).

Table 2.5 Five domesticated *Capsicum* species with their distinct characteristics, according to Eshbaugh.

Taxa	Chili fruit types	Characteristics	Distribution
C. annuum var. *annuum* L.	Jalapino, pablano, Anaheim, ancho, bell, cayenne, Serrano	Flowers with white to bluish white petal, single; calyx teeth lacking or short; non-prominent constriction between calyx base and pedicel	Mexico
C. chinense Jacq.	Habanero, Scotch bonnet, rocotillo	Flowers 2 or more/node, pendant, calyx teeth lacking, dull white but rarely greenish white petal; prominent constriction between calyx base and pedicel; anthers blue to violet, rarely yellow; seed cream to yellow	Amazonas
C. frutescens L.	Tabasco, malagueta, African Birdseye, piri-piri, Thai khee noo suan	Flowers 2 or more/node, greenish white; non-prominent constriction between calyx base and pedicel; calyx teeth absent; anthers blue to violet, rarely yellow; seed cream to yellow	Caribbean
C. baccatum var. *pendulum* (Willd.) Eshbaugh	Aji, aji amarillo, cuero de oro, cumbai	Flowers cream to white petal with yellow or green markings; seed cream	Peru
C. pubescens Ruiz & Pavon	Rocoto, locoto, chile manzana	Flowers purple or white petals with 5-8 lobes; seed dark brown or black; leaves large, rugose, pubescent	Mid-elevated Andes (1500–3000 m)

Source: data from Eshbaugh, W.H., *Peppers: Botany, Production and Uses*, CAB International, Wallingford, UK, 2012.

Table 2.6 *Capsicum* species accepted by the Plant List (2013)

Taxa	Distribution[a]	*n*
1. *C. annuum* L. var. *annuum* L. var. *glabriusculum* (Dunal) Heiser & Pickersgill	Colombia, north to southern United States	12, 24
2. *C. baccatum* L. var. pendulum (Willd.) Eshburg var. *praetermissum* (Heiser & P.G.Sm.) Hunz.	Argentina, Bolivia. Brazil, Paraguay, Peru	12
3. *C. buforum* Hunz.	Brazil (e)	13
4. *C. caballeroi* M. Nee	Bolivia (e)	
5. *C. campylopodium* Sendtn.	Southern Brazil, Paraguay (e)	13
6. *C. cardenasii* Heiser & Smith	Bolivia (e)	12
7. *C. ceratocalyx* M.Nee	Bolivia (e)	
8. *C. chacoense* Hunz.	Argentina, Bolivia	12
9. *C. chinense* Jacq.	South America, Amazonas, Caribbean	12
10. *C. coccineum* (Rusby) Hunz.	Bolivia, Peru	
11. *C. cornutum* (Hiern) Hunz.	Southern Brazil (e)	13
12. *C. dimorphum* (Miers) Kuntze	Colombia	
13. *C. dusenii* Bitter	Southeastern Brazil (e)	
14. *C. eshbaughii* Barboza	Bolivia (e)	12
15. *C. eximium* Hunz.	Argentina, Bolivia	12
16. *C. flexuosum* Sendtn.	Brazil, Paraguay	12
17. *C. friburgense* Bianch. & Barboza	Southern Brazil (e)	13
18. *C. galapagoense* Hunz.	Ecuador (e)	12
19. *C. geminifolium* (Dammer) Hunz.	Colombia, Ecuador	
20. *C. havanense* Kunth	Cuba[b]	
21. *C. hookerianum* (Miers) Kuntze	Ecuador (e)	
22. *C. hunzikerianum* Barboza & Bianch.	Southern Brazil (e)	
23. *C. lanceolatum* (Greenm.) C.V.Morton & Standl.	Mexico, Guatemala, Honduras	13
24. *C. leptopodum* (Dunal) Kuntze	Brazil (e)	
25. *C. lycianthoides* Bitter	Ecuador, Colombia[b]	
26. *C. minutiflorum* (Rusby) Hunz.	Argentina, Bolivia, Paraguay	
27. *C. mirabile* Sendtn.	Southern Brazil (e)	13
28. *C. mositicum* Toledo ex Handro	Brazil[b]	
29. *C. parvifolium* Sendtn.	Colombia, northern Brazil, Paraguay	12

(Continued)

Table 2.6 (Continued) *Capsicum* species accepted by the Plant List (2013)

Taxa	Distribution[a]	*n*
30. *C. pereirae* Barboza & Bianch.	Southern Brazil (e)	13
31. *C. pubescens* Ruiz & Pav.	Andean South America, Guatemala, Mexico	12
32. *C. ramosissimum* Witasek	Brazil[b,c]	
33. *C. recurvatum* Witasek	Southern Brazil (e)	
34. *C. rhomboideum* (Dunal) Kuntze	Southern Mexico to Peru and Venezuela	13
35. *C. schottianum* Sendtn.	Argentina, southern Brazil, southeastern Paraguay	13
36. *C. scolnikianum* Hunz.	Peru (e)	
37. *C. sinense* Jacq.	Peru[b]	12
38. *C. spina-alba* (Dunal) Kuntze	Brazil[b]	
39. *C. tovarii* Eshbaugh, P.G.Sm. & Nickrent	Peru (e)	12
40. *C. villosum* Sendtn.	Southern Brazil (e)	13

[a] Distribution data without superscript from Eshbaugh (2012).
[b] Tropicos® (2017)= tropicos.org.
[c] IPNI (2012) = The International Plant Names Index.
e, endemic according to Eshbaugh (2012); *n*, haploid chromosome number according to Eshbaugh (2012).

1. Style cylindrical, equal in width from the base to the apex. Fruit red, generally elliptical, ovoid, or sometimes globose. Seed yellowish-brown, the episperm smooth. Corolla 4–7.5 mm long.
    2. Corolla stellate, unspotted, white or cream-colored, lobes generally oblong and more or less the same length as the limb and tube. Filaments as long as or shorter than the anthers. Northern and northeastern Brazil (Acre, Amazonas, Maranhão, Rondonia, Roraima)....................... ***C. annuum* var. *glabriusculum* (Dunal) Heiser & Pickersgill**
    2. Corolla rotate, white with greenish-yellow spots on the lobes and limb inside, the lobes broader than long and markedly shorter than the limb and tube. Filaments generally 1.5 times or more longer than the anthers.
        3. Corolla with the inside margin white. South and southeastern Brazil (Espírito Santo, Mato Grosso do Sul, Minas Gerais, Paraná, Rio de Janeiro, Rio Grande do Sul, Santa Catarina, São Paulo)......... ***C. baccatum* L. var. *baccatum***
        3. Corolla with the inside margin lilac or violet. Southeastern and west-central Brazil (Goiás, Minas Gerais, Paraná, Santa Catarina, São Paulo) ....***C. baccatum* var. *praetermissum* (Heiser & Smith) Hunz**

1. Style clavate, widening from a moderately narrow base to a gradually broadened apex. Fruit generally yellow or yellowish-green at maturity, rarely red-colored, globose or globose-depressed or globose-compressed. Seed generally brownish or blackish (yellowish-brown only in *C. parvifolium*), the episperm foveolate with spine-like projections. Corolla (5.5) 6–15 (16) mm long.
    4. Corolla entirely pink or lilac, clearly campanulate to urceolate, tube (5.5) 7–9 (11) mm. Leaves generally ovate. Eastern Brazil (Rio de Janeiro) .... ***C. friburgense* Bianchetti & Barboza**
    4. Corolla white with yellowish-green and sometimes also purple spots inside, stellate or rotate, never campanulate-urceolate, tube (2) 2.6–6 (8) mm. Leaves ovate, elliptical or narrowly elliptical.
        5. Pedicels non-geniculate at anthesis, the flowers pendant.
            6. Shrubs or trees up to 4 m tall or more. Fascicles 5–20-flowered. Calyx 5-toothed. Anthers as long as or longer than the filaments. Seed yellowish-brown. Northeastern Brazil (Bahía, Ceará, Paraiba, Pernambuco, Piauí, Rio Grande do Norte) .... ***C. parvifolium* Sendtn**
            6. Shrubs 0.5–2 (3) m tall. Flowers solitary or the fascicles 2–3-flowered. Calyx toothless or with 5 minuscule teeth. Anthers clearly shorter than the filaments. Seeds brownish or blackish.
                7. Corolla white with yellowish-green spots in the lobes and limb inside, 5.5–6 mm long. Leaves membranaceous, ovate, 2–3 (3.5) times longer than broad, glabrescent to pubescent. Fruits red at maturity. South and southeastern Brazil (Minas Gerais, Paraná, Rio Grande do Sul, São Paulo, Santa Catarina) .... ***C. flexuosum* Sendtn**
                7. Corolla white with purple spots followed by an interrupted yellowish-green zone in the lobes and limb, 9–10 mm long. Leaves coriaceous, elliptical to narrowly elliptical, 3–5.5 (10) times longer than broad, glabrate. Fruits yellowish-green at maturity. Southeastern Brazil (Espírito Santo, Minas Gerais)...................... ***C. pereirae* Barboza & Bianchetti**
        5. Pedicels geniculate at anthesis, the flowers twisted 90°.
            8. Corolla lacking purple spots inside.
                9. Calyx toothless. Corolla with yellow or golden spots in lobes and limb. Ovules 2 per locule. Androecium

heterodynamous with 3 short stamens and 2 long stamens. Fruit globose-compressed. Southeastern Brazil (Espírito Santo, Minas Gerais, Rio de Janeiro) ....*C. campylopodium* **Sendtn**

9. Calyx with 5, or 6 to 9 horizontal or recurved teeth. Corolla with greenish spots inside. Ovules 5–8 per locule. Androecium homodynamous with all stamens equal in length. Fruits globose-depressed. South and southeastern Brazil (Paraná, Rio de Janeiro, Santa Catarina, São Paulo) .... *C. recurvatum* **Witas**

8. Corolla with purple or brownish or violaceous spots followed by yellowish-green zones inside.
   10. Calyx toothless or sometimes with 5 tiny teeth. Southeastern Brazil (Minas Gerais, Rio de Janeiro, São Paulo) .... *C. schottianum* **Sendtn**
   10. Calyx 5–10-toothed.
       11. Calyx with only 5 short teeth (0.5–3 mm long).
           12. Plants glabrescent, the hairs antrorse. Leaves elliptical to narrowly elliptical. Southeastern Brazil (Minas Gerais, Rio de Janeiro, São Paulo) .... *C. mirabile* **Mart**
           12. Plants densely hairy, the hairs flexuous and patent on stems, petioles, pedicels, and sometimes also on the leaf nerves beneath. Leaves ovate. Southeastern Brazil (Minas Gerais, Rio de Janeiro, São Paulo) .... *C. villosum* **Sendtn**
       11. Calyx up to 6–10 long teeth (3.2–6 mm long).
           13. Shrubs 1.2–1.8 m tall, densely hairy. Corolla (8) 9–12 (14) mm long. Leaves membranaceous, ovate to broadly ovate. Southeastern Brazil (Rio de Janeiro, São Paulo) .... *C. cornutum* **(Hiern) Hunz**
           13. Shrubs up to 3 m tall, glabrate. Corolla 10–14 (16) mm long. Leaves coriaceous, slightly ovate to elliptical. Southeastern Brazil (São Paulo) .... *C. hunzikerianum* **Barboza & Bianchetti**

2.3 Capsicum *evolution, species relationships, and cytogenetics*

Phylogenic relationships of 11 *Capsicum* species were studied using DNA sequences from two non-coding regions including chloroplast stpB-rbcL spacer and nuclear waxy introns (Walsh & Hoot 2001). *C. ciliatum* (syn. *rhomboideum*), *C. eximium*, and *C. tovarii* were the most divergent species within *Capsicum*. *C. ciliatum* was isolated from the rest of the *Capsicum*. The *Capsicum* species were grouped corresponding to the known complexes, i.e., *C. annuum/chinense/frutescens*, but included *C. galapagoense*, *C. eximium/cardenasii*, and *C. baccatum/chacoense* (Figure 2.12).

Capsicum evolution, based on cytogenetic study, was summarized by Moscone et al. (2007). The *Capsicum* species were distinguished into two groups based on chromosome numbers, $2n=24$ and $2n=26$. The 26 chromosome number appeared to be the most common of the wild species, except for *C. flexuosum* and *C. parvifolium* (Pozzobon et al. 2006). *C. chacoense* appeared to be the most primitive taxon, and the species with chromosome number 26 were likely to be more advanced than the species with $2n=24$. Most $2n=26$ *Capsicum* species are local endemics found in southeastern Brazil. The wild $2n=24$ are found elsewhere throughout the range of *Capsicum* in South America (Eshbaugh 2012). An exception is for one accession of *C. annuum* var. *glabriusculum* with tetraploid $2n=4x=48$, which was reported by Pickersgill (1977). More recently, a tetraploid cultivar of *C. annuum* was identified from South Sikkim in the eastern Himalayas of India (Jha et al. 2012). Meiotic analysis suggested that the tetraploid cultivar was not autotetraploid or due to self-duplication. The *Capsicum* species that are more advanced in evolution have a higher DNA content (1C or haploid DNA content 3.35–5.77 pg) and heterochromatin amount (1.8%–38.9% of the karyotype length) (Pickersgill 1977).

A possible evolutionary relationship among the *Capsicum* species based on karyotype features was proposed by Moscone et al. (2007) and is displayed in Figure 2.13. All *Capsicum* species share a common ancestor, which was a diploid with $x=12$. *C. chacoense* diverged at the earliest stage of the evolution line, and the *C. annuum* complex (*C. annuum*, *C. chinense*, and *C. frutescens*) evolved soon after. The primitive group, which is represented by species with white flowers, is *C. galapagoense*, *C. rhomboideum*, and *C. parvifolium*. The recent species group started with *C. baccatum* and *C. praetermissum*, and the three most advanced groups were beyond this point. The first was the species group with purple flowers (*C. eximium*, *C. cardenasii*, *C. pubescens*, and *C. tovarii*), the second was *C. flexuosum*, and the third was the most advanced (the majority were $2n=26$) group with *C. mirabile* as the core species.

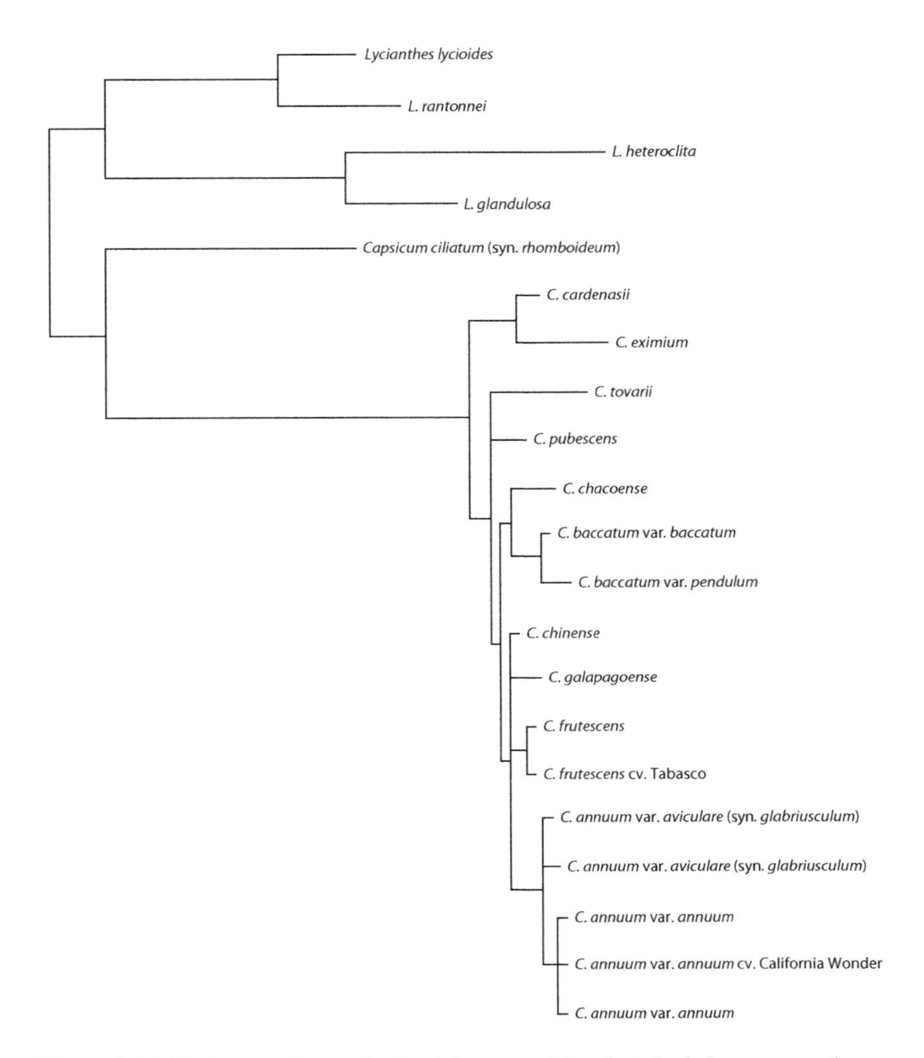

Figure 2.12 Phylogenetic tree derived from combined *atpB-rbcL* spacer and *waxy* data of *Capsicum* spp. (Redrawn and modified from Walsh, B.M., Hoot, S.B., *IJPS*, 162, 1409–1418, 2001.)

Another model of *Capsicum* evolution was proposed by Eshbaugh (2012). This model included three independent evolutionary lines: *C. baccatum* var. *pendulum*, which arose from an ancestral wild *C. baccatum* var. *baccatum*; *C. pubescens* from another ancestor in the *C. eximium* gene pool; and third, the three domesticated taxa arose from a common ancestor (Figure 2.14).

More recent molecular studies suggested different relationships among *Capsicum* species, with *C. chinense* and *C. frutescens* being closely

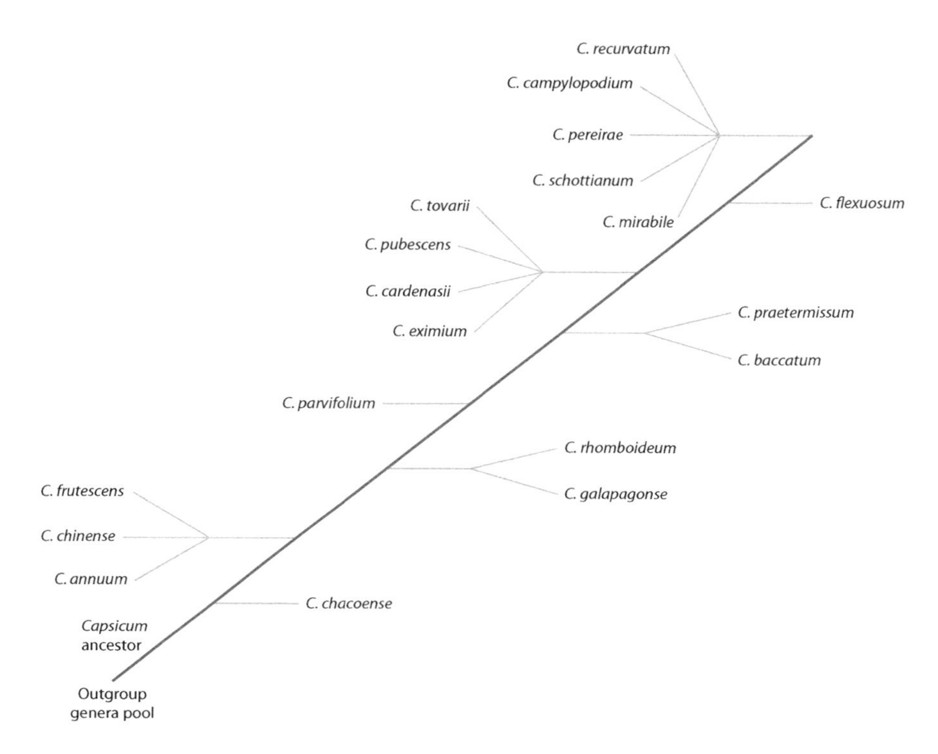

Figure 2.13 Possible evolutionary relationships among *Capsicum* species based on karyotype features. Outgroup genera pool ($2n=24$): *Lycianthes, Dunalia, Vassobia, and Eriolarynx*, proposed by Moscone et al. (2007). (Redrawn from Moscone, E.A. et al., *Acta Hortic.*, 745, 137–170, 2007.)

related (Ibiza et al. 2012; Gonzalez-Perez et al. 2014), which was in agreement with previous reports. *C. cardenasii* and *C. eximium* were single species. Although *C. baccatum* and *C. praetermissum* formed a compact group, *C. praetermissum* is now accepted as a variety of *C. baccatum* (the Plant List 2013). Interestingly, the *C. chacoense* was close to the *C. baccatum* complex, but *C. tovarii* was a separate species (Figure 2.15; Ibiza et al. 2012; Gonzalez-Perez et al. 2014).

When all the reports are combined, *Capsicum* species relationships are still debatable according to the different findings based on cytogenetic and molecular analyses. Walsh and Hoot (2001) summarized the crossability of wide hybridization between different *Capsicum* species from various studies during 1948–1976 (Figure 2.16). Interestingly, the only highly fertile hybrids were achieved from *C. eximium* and *C. cardenasii*. *C. frutescens* appeared to be closer to *C. chinense* than to *C. annuum*.

Cytogenetics: The $2n=24$ species have comparatively symmetrical karyotypes, mostly with 11 metacentric (m) and 1 subtelocentric (st) or

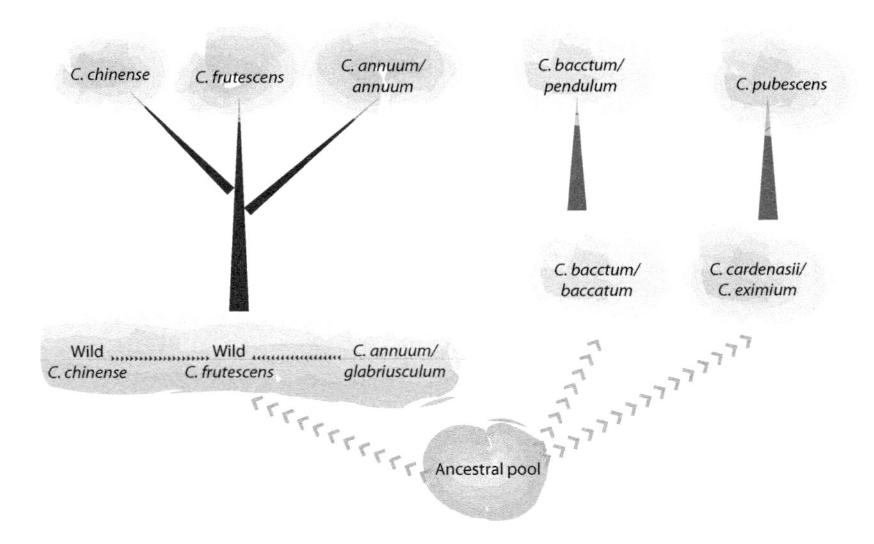

Figure 2.14 Hypothetical evolutionary model of the five domesticated *Capsicum* species, proposed by Eshbaugh (2012). (Redrawn from Eshbaugh, W.H., *Peppers: Botany, Production and Uses*, CAB International, Wallingford, UK, 2012.)

submetacentric (sm). The $2n=26$ species have more asymmetrical complements with more sm and st. *Capsicum* karyotypes of six wild and five cultivated species were physically mapped with 5S rDNA locus using fluorescent *in situ* hybridization (FISH; Aguilera et al. 2016). The 5S rDNA genes are tandem repeats ranging from thousands to hundreds of thousands of copies in a genome, which can be useful for the identification of individual chromosomes. All *Capsicum* species have one locus of 5S rDNA located on the short arm (p) of different chromosomes (Figure 2.17; Table 2.7). Other classes of rDNA, including 18S and 45S, showed variations in number, locations, and size between different species, varieties, and cultivars of *Capsicum* (Moscone et al. 2011; Cruz et al. 2017).

Based on the metaphase cell, the chromosome structure contains a constriction point, the centromere, which divides the chromosome into two arms, p and q. The short arm refers to *p*, and therefore, the long arm refers to *q*. There are four types of chromosome, identified by the different positions of the centromere on the chromosomes; a metacentric chromosome (m) has its centromere in the center, such that p equals q; a telocentric chromosome (t) has its centromere positioned at one end of the chromosome, such that the chromosome has only one arm; a submetacentric chromosome (sm) has the centromere located near the center, such that p is shorter than q; a subtelocentric (st) or acrocentric chromosome has the centromere located toward one end of the chromosome, such that p is very much shorter than q (Naranjo et al. 1983).

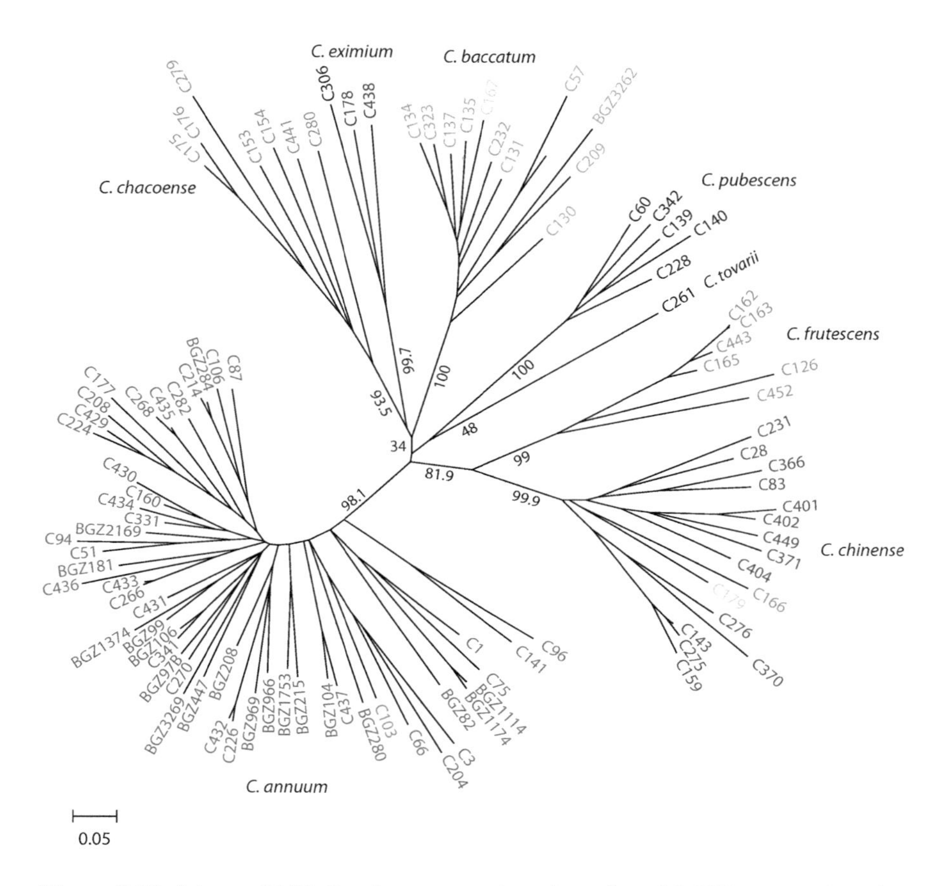

Figure 2.15 A tree of 102 *Capsicum* accessions based on 39 SSRs, grouping the accessions according to species. (Modified from Gonzalez-Perez, S., et al., *PLoS ONE*, 9, e116276, 2014. Figure licensed under the Creative Commons Attribution 4.0.)

2.4 Capsicum *germplasm genebanks and resource management*

Wild crop relatives are vital sources of genetic variation for improving domesticated species (Schoen & Brown 1993). The diversity of alleles in crops makes them vulnerable to loss due to the reduction of natural population sizes. Wild species are found in nature, and many of them have been dramatically threatened by habitat reduction. Globally, the conservation of crop germplasm has been taken very seriously since 1974, when the International Board of Plant Genetic Resources (IBPGR) was formed under the aegis of the Consultative Group on International Agricultural Research (CGIAR). The IBPGR's mandate is to promote an international

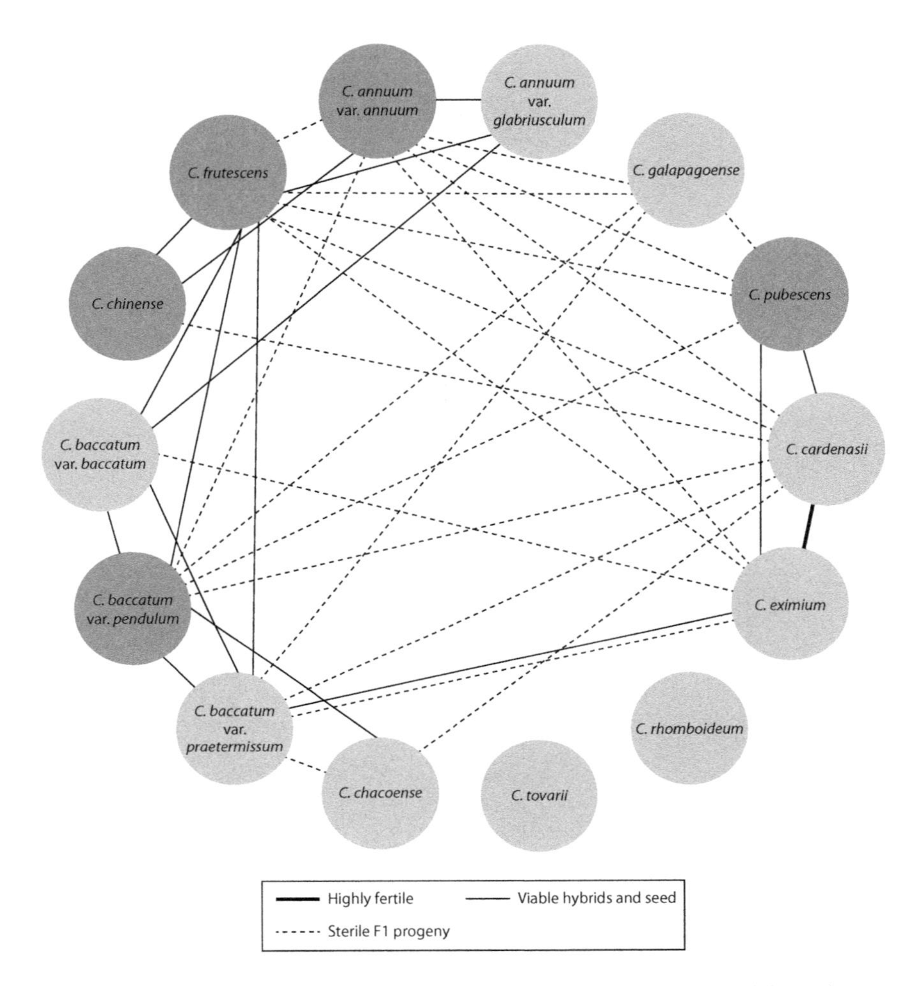

Figure 2.16 Crossability of *Capsicum* species summarized by Walsh and Hoot (2001). *C. annuum* var. *glabriusculum* is synonym of *C. annuum* var. *aviculare* and *C. rhomboideum* is synonym of *C. ciliatum*. Dark gray circles are cultivated species, and light gray circles are wild species. (Modified and redrawn from Walsh, B.M., Hoot, S.B., *IJPS*, 162, 1409–1418, 2001.)

network of genetic resource centers to further the collection, conservation, documentation, evaluation, and use of plant germplasm. As a result, it hopes to contribute to raising the standard of living and welfare of people throughout the world. IBPGR was renamed the International Plant Genetic Resources Institute (IPGRI) in 1991. IPGRI and the International Network for Improvement of Banana and Plantain (INIBAP) became a single organization in 2006, named Bioversity International. Bioversity has worldwide partnerships.

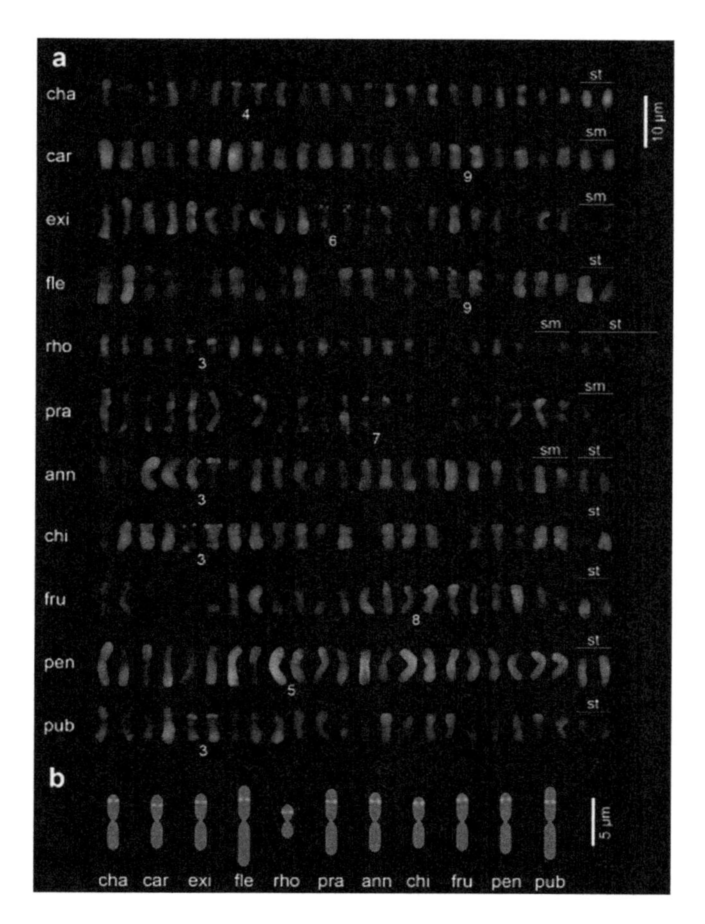

Figure 2.17 Karyograms of wild and cultivated *Capsicum* species, derived from fluorochrome (blue)-stained somatic metaphase chromosomes showing number and position of 5S rDNA locus (green fluorescence). Chromosome structures: sm is submetacentric, st is subtelocentric. (Source: Aguilera, P.M. et al., *An. Acad. Bras. Ciênc.*, 88, 117–125, 2016. Figure licensed under the Creative Commons Attribution 4.0.)

Genebanks around the world have important duties to collect, conserve, and evaluate crop germplasm in their collections so as to ensure that the evaluated germplasm can be used. Germplasm management has become problematic for genebanks worldwide due to the growing size of the germplasm collections. Average annual costs of *ex situ* crop germplasm conservation at the World Vegetable Center (WVC) were estimated to be US$10.08 per accession (Schreinemachers et al. 2014). The WVC holds over 67,000 accessions, making it the world's fifth largest international public genebank and the largest for vegetable germplasm. Its average annual operating cost is US$0.684 million, of which 74% is used for labor and 17% for storage costs.

Table 2.7 Chromosome features of wild (w) and cultivated (c) *Capsicum* species with position of 5S rDNA

Taxa	Abbreviation	2*n*	Karyotype formula	Number of loci	Position
C. chacoense (w)	cha	24	22 m + 2 st	1	4p
C. cardenasii (w)	car	24	22 m + 2 sm	1	9p
C. eximium (w)	exi	24	22 m + 2 sm	1	6p
C. flexuosum (w)	fle	24	22 m + 2 st	1	9p
C. rhomboideum (w)	rho	26	20 m + 2 sm + 4 st	1	3p
C. baccatum var. *praetermissum* (w)	pra	24	22 m + 2 sm	1	7p
C. annuum var. *annuum* (c)	ann	24	20 m + 2 sm + 2 st	1	3p
C. chinense (c)	chi	24	22 m + 2 st	1	3p
C. frutescens (c)	fru	24	22 m + 2 st	1	8p
C. baccatum var. *pendulum* (c)	pen	24	22 m + 2 st	1	5p
C. pubescens (c)	pub	24	22 m + 2 st	1	3p

Source: Modified from Aguilera, P.M. et al., *An. Acad. Bras. Ciênc.*, 88, 117–125, 2016.
2*n*, somatic chromosome number; p, short chromosome arm (of a certain chromosome). Karyotype formula: m, metacentric; sm, submetacentric; st, subtelocentric.

2.4.1 *Core collection*

A large collection size can deter use; thus, a core collection concept was proposed (Frankel 1984; Frankel & Brown 1984), in which a limited-size subset of an existing collection would be established with minimal similarities between its entries. As a result, the core collection represents the genetic diversity of the entire collection, which has been suggested to compose approximately 10% of the total number (Brown 1989). The selection criteria for the core members should be considered from available data on geographic origin, genetic characteristics, and the possible value to plant breeders and other users. Therefore, a good core collection should (1) contain no redundant entries, (2) have authenticity of origin, (3) be sufficiently large for reliable conclusions as a whole collection, (4) be able to predict sources of useful variation, (5) represent major subspecific taxa and geographic regions, (6) represent more broadly adapted rather than intensely specialized alleles, and (7) contain maximum diversity.

Core collection has become accepted as an efficient tool for improving the conservation and use of germplasm, which was recommended as an essential activity to improve plant genetic resources in the Global Plan of Action for the Conservation and Sustainable Utilization of Plant Genetic resources for the Food and Agriculture Organization (FAO) in 1996 (van Hintum et al. 2000). Establishing a core collection comprises five steps, as shown in the following subsections.

- Identifying materials to be represented

  Materials to be represented by the core differ case by case; most are materials of a certain crop in a genebank collection.

- Size of core collection

  Theoretically, a core collection is limited to 10% of the entire collection and always less than 2,000 entries. Practically, most core collections are between 5% and 20% and reach up to 2,000 accessions. When the core is developed from a very large collection, the core size may be much smaller than 5%. The international barley core collection, for example, contains 1,600 accessions—equivalent to less than 0.3% of the world barley collection.

- Dividing into genetically distinct groups

  The effective representation of a whole collection by a core depends on meaningful groupings of germplasm. The strategy used in this case is stratification, which constructs groups to maximize variation between groups and minimize variation within groups. This step is the key to developing a good core collection. A hierarchical procedure is usually practiced, first making major divisions and subsequently smaller subgroups. Often, the first division is based on taxonomy by separating the wild from the domesticated species. Subsequently, within these groups, the species and then the subspecies are identified. A structured hierarchy can be developed by combining taxonomy and knowledge of domestication, distribution, breeding history, cropping pattern, and use to form a diversity tree. The following table shows an example of a structured diversity tree for the *Lactuca* collection at the Center for Genetic Resources, the Netherlands (CGN): total 2,332 accessions (van Hintum et al. 2000).

Wild *Lactuca*	Domesticated *Lactuca*
598 primary genepool	743 butterhead lettuce
169 secondary genepool	225 cos lettuce
55 tertiary genepool	237 crisp lettuce
2 unknown	163 cutting lettuce
	61 latin lettuce
	32 stalk lettuce
	16 oilseed lettuce
	32 unknown

- Number of entries per group

After the groupings are defined, the number of entries from each group has to be decided. Three basic approaches have been used, based on number of accessions, molecular marker diversity, and informal knowledge of users or curators. The three strategies, constant, proportional, and logarithmic, are based on group size and were originally proposed by Brown (1989). The *constant* method allocates an equal number of entries to each group, no matter what the number of accessions in the group. The *proportional* method allocates entries to a group in proportion to the number of accessions in each group. The *logarithmic* method allocates entries to each group in proportion to the logarithm of the number of accessions in the group. These three methods have been compared, and the outcome favors the proportional and logarithmic methods. However, a recommendation is made to include at least one accession from each group, no matter which method is used and how small the group.

Based on marker density, the crop diversity is estimated from the allele frequency (h index) or number of distinct allele types (range of marker loci in each accession). The H or heterozygosity method is aimed at maximizing allele richness in the core. The alternative M method uses a linear program to search for the combination of accessions that will maximize the number of observed alleles at the marker loci of the core while keeping the total number of samples to a specified limit and ensuring a minimum number from every group. This method guarantees the maximum allele richness for the marker loci used and directly takes the amount of variation and the divergence pattern at those loci. Therefore, the M method not only determines the accession number from different groups but also identifies the accessions to be included. The PowerCore computer software was developed (Kim et al. 2007a) to help select germplasm based on the M strategy. However, the effectiveness of the different methods depends on the meaningful groupings of the germplasm.

A study by Mongkolporn et al. (2015) applied PowerCore to select chili accessions to form a core collection, which reduced the collection size down to 12% (28 were selected from 230 accessions). A dendrogram generated from 10 anchored SSR loci shows the relatedness of the chili germplasms, which are divided into two main groups with several subgroups (Figure 2.18). The study proved that the PowerCore program could maintain maximum allele richness (42 SSR alleles) as effectively as the whole chili collection; however, the 28 accessions selected by the program did not cover all the chili subgroups, which could be improved by manual selection from unselected subgroups, if necessary. Informal knowledge can be applied to determine the number of entries in a group and to adjust the numbers obtained using other methods. Adjustments can be made based on reasons such as accessions that are important to the user community or that are not interesting. Therefore, such reasons can affect the group size either positively or negatively.

- Choosing the entries

The final step in establishing a core collection is the selection of entries that will best represent the group and best serve the function

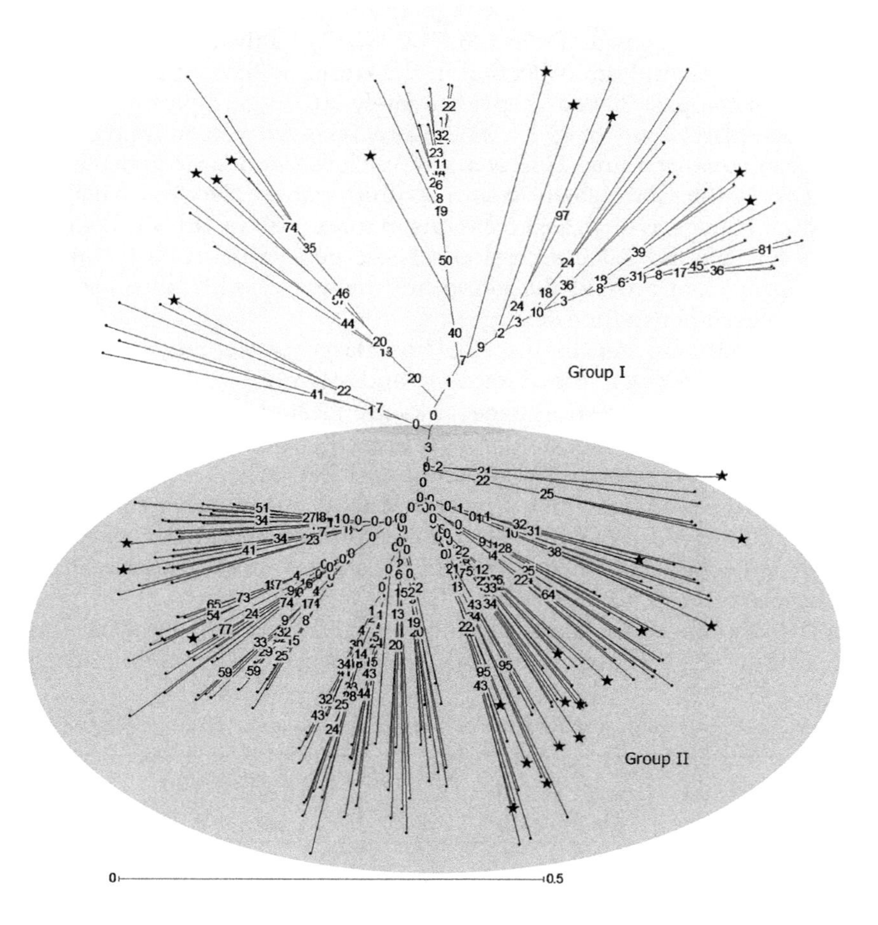

Figure 2.18 Dendrogram illustrating the relatedness of 230 *Capsicum* accessions generated from 10 anchored SSR loci. The asterisks represent the 28 chili accessions selected by the PowerCore software version 1.0 to form a core collection. (Dendrogram courtesy of Soontree Hanyong.)

and purpose of the core. The fastest and easiest way to select the entries is at random, avoiding the selection of a set of sequential accession numbers, which could be biased. Often, blocks of accessions of similar origin enter a collection together and are given sequential numbers. Moreover, additional data from the marker, characterization, evaluation, or passport of accessions should be combined for consideration.

2.4.2 Capsicum *germplasm genebanks*

There are three major international genebanks that have large collections of *Capsicum* germplasm (Table 2.8). The WVC in Taiwan holds the largest *Capsicum* germplasm collection in the world, with 8,165 accessions in total. Over the past 30 years, approximately 30,000 germplasm materials have been distributed, of which 20% are genebank accessions and 80% are improved advanced lines (Lin et al. 2013). The second large genebank for *Capsicum* is the Germplasm Resources Information Network (GRIN) in the United States, with 6,266 accessions in total, and the third is the CGN, which has around 1,000 accessions. These genebanks collect *Capsicum* germplasm from around the world and have accessible databases and available germplasm for exchange.

Some national genebanks also hold large collections of *Capsicum* germplasm. However, the databases and germplasm exchanges may not be available. Interesting genebanks are located in Peru and Bolivia, where the primary center of *Capsicum* diversity exists. The Peruvian collection has 712 accessions, of which almost half are *C. pubescens*, while the Bolivian has 487, of which almost half are *C. baccatum* (van Zonneveld et al. 2015). *C. chinense* is the second most numerous in both collections. Brazil's collection has a rich variety of *C. chinense* and *C. frutescens* germplasm (Carvalho et al. 2014, 2017).

The two major *Capsicum* producers of the world are in Asia: India and China. The Chinese genebank holds largely *C. annuum* landraces

Table 2.8 *Capsicum* germplasm collections of the key international genebanks

Gene	Total	Major species component						
bank	number	Ca	Cb	Cc	Cf	Cp	Wild	Unknown
WVC	8165	5377	383	499	703	31	45	1127
GRIN	6266	3410	382	491	282	45	na	na
CGN	1011	783	na	116	45	na	56	205

Genebanks: GRIN, Germplasm Resources Information Network, United States; CGN, Center of Genetic Resources, the Netherlands; WVC, the World Vegetable Center, Taiwan.
Species component: Ca, *C. annuum*; Cb, *C. baccatum*; Cc, *C. chinense*; Cf, *C. frutescens*; Cp, *C. pubescens*; wild, wild species; na, not available.

(Zhang et al. 2016). Indian landraces belong to *C. annuum*, *C. chinense*, and *C. frutescens* (Yumnam et al. 2012). Thailand also has a large collection of *Capsicum* germplasm (approximately 3000 accessions) at Kasetsart University, Tropical Vegetable Research Center (Mongkolporn & Taylor 2011; Mongkolporn et al. 2015). Within the collection, 250 are Thai chili landraces that belong to *C. annuum* and *C. frutescens*.

2.5 Conclusions and remarks

Capsicum is one of the world's top-ranked vegetables and spices. It has a long and interesting history and is one of the most genetically diverse crops in the world. Originating in South America, *Capsicum* has spread around the world. To date, there are 40 accepted species, with some species still debatable and awaiting acceptance. The *Capsicum* genome, including chloroplast and mitochondria genomes, has been sequenced, which was a significant event that allows the discovery of important genes. However, no matter how much progress the technology makes, crop diversity is still essential and needs to be enhanced or at least maintained, since the germplasm is the only source of important genes for the future. Fortunately, there are key international genebanks that collect, conserve, evaluate, and exchange the germplasms of crops, including but not limited to *Capsicum*.

Colorful and diverse *Capsicum* spp.

Scoville unit	Chili type
15,000,000	Pure capsaicin
9,000,000	Nordihydrocapsaicin
2,000,000–5,500,000	US chili spray
700,000–1,000,000	Bhut jolokia
350,000–600,000	Red savina habanero
100,000–350,000	Habanero, scotch bonnet
100,000–200,000	Jamaican, rocoto
50,000–100,000	Thai, chiltepin
30,000–50,000	Cayenne, tabasco
10,000–25,000	Serrano
5,000–10,000	Wax
2,500–8,000	Jalapeno
1,500–2,500	Rocotillo
1,000–1,500	Poblano
500–2,500	Anaheim
100–500	Pimento
0	Bell

Figure 1.4 Scoville heat units of diverse chili types. (Redrawn from O'Neill, J. et al., *Pharmacological Reviews*, 64, 939–971, 2012.)

Table 1.3 List of *Capsicum annuum*

	Bell: non-pungent, blocky type, 10 cm long. Bell cultivars have ripe fruit in various shades of yellow, orange, red, and brown. Uses: fresh and cooking.
	Pimento or pimiento: non-pungent, heart shaped or round, thick wall, red ripe fruit, 7–10 cm long and 5–7 cm wide. Uses: processed foods such as pimento cheese, stuffed olives, or fresh and roasted.
	Wax: non-pungent to mildly pungent, yellow immature fruit turning orange or red when ripe; well known as the long fruit Hungarian wax or banana, 10 cm long and 4 cm wide. Short type is 5 × 2 cm. Uses: pickles and fresh.
	Ancho: mildly pungent, heart shaped, pointed, thin wall, dark green when immature and red or chocolate brown when ripe, 8–15 cm long; growth is restricted to Mexico. Uses: stuffed foods.
	Pasilla: mildly pungent, long and slender fruit, 15–30 cm long and 2.5–5 cm wide; ripe fruit is brown. Uses: dried to make sauce.
	Cayenne: very pungent, long and wrinkled red ripe fruit, 13–25 cm long and 1.2–2.5 cm wide, widely grown around the world. Uses: dried, ground, and sauces.
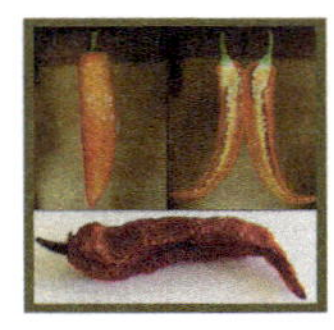	Mirasol: moderately pungent, conical shape, erect fruit has slight curve, 7–10 × 1–2 cm, thin wall, becomes translucent when dried. Uses: dried. (Photographs courtesy of Bruno B. Defilippi and Pitchayapa Mahasuk.)
	Pepperoncini: non-pungent to mildly pungent, large fruit, 7.5–12.5 cm long, irregular, thin wall. Uses: pickles.

Table 1.3 (Continued)

	Jalapeno: very pungent, conical fruit, thick wall, dark green immature fruit, 5–10 cm long and 2.5–3.8 cm wide. Dry fruit skin showing netting pattern or corkiness. Uses: canned, pickles, and dried.
	Serrano: very pungent, cylindrical fruit, 5–10 cm long and 1 cm wide, medium-thick wall, no corkiness. Uses: salsa sauce.
	De Arbol: moderately pungent, long and slender fruit, 5–8 cm long and 0.5–1 cm wide, fruit translucent when dried. Uses: dried.
	Cherry: non-pungent to mildly pungent, small round fruit around 2.5 cm. Uses: fresh and pickles.
	Thai large bird or Khee Noo: very pungent, long fruit, 3–12 cm long. Uses: fresh, dried, ground, and chili paste.
	Thai Chee Faa: mildly to moderately pungent, 5–20 cm long fruit, similar to Cayenne but with smooth skin, color yellow to red. *Chee Faa* is Thai for "pointing up to the sky"; however, Chee Faa is not erect. Uses: fresh, dried, ground, chili paste, and sauces.

Sources: Data from Bosland, P., Votava, E.J., *Peppers: Vegetable and Spice* Capsicums, CAB International, Reading, UK, 2012; www.cayennediane.com, 2017; www.chilipepper-madness.com, 2017.

Table 1.4 List of *Capsicum chinense*

	Habanero: very pungent, lantern shaped, orange or red ripe fruit, 6 cm long and 2.5 cm wide. Uses: fresh, salsa sauce, fermented to make spicy sauce.
	Bhut Jolokia: extremely pungent, lantern shaped, orange or red ripe fruit, 5–7.5 cm long. Uses: defense products.
	Aji dulce: very pungent, fruit size similar to Habanero, red, orange, or yellow fruit, 2.5–5 cm long and 2.5–3.2 cm wide. Uses: fresh, salsa sauce, and fermented to make spicy sauce.
	Charapita: very pungent, very small fruit with 0.6 cm diameter, round and thin flesh, erect, yellow or red ripe fruit. Uses: cooking.
	Biquinho or Chupentinho: moderately pungent, small round fruit around 2.5 cm long with beak-shaped end, bright yellow or red ripe fruit. Uses: ornament, pickles, and cooking.

Sources: Data from Bosland, P., Votava, E.J., *Peppers: Vegetable and Spice* Capsicums, CAB International, Reading, UK, 2012; www.cayennediane.com, 2017; www.chilipepper-madness.com, 2017.

Table 1.5 List of *Capsicum frutescens*

	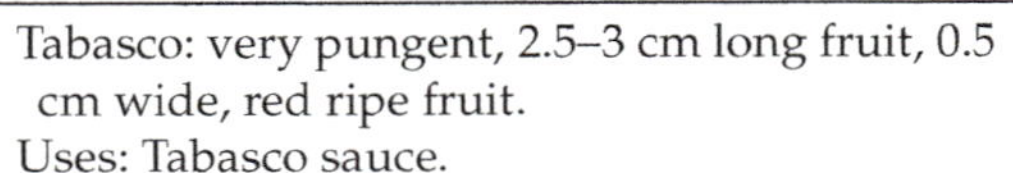 Tabasco: very pungent, 2.5–3 cm long fruit, 0.5 cm wide, red ripe fruit. Uses: Tabasco sauce.
	Thai small bird: moderately to very pungent, fruit length <3 cm, erect, white to green immature, red ripe fruit, unique aroma. Uses: fresh, dried, ground, and chili paste. (Photograph courtesy of Kietsuda Luengwilai.)

Source: Data from Bosland, P., Votava, E.J., *Peppers: Vegetable and Spice* Capsicums, CAB International, Reading, UK, 2012.

Table 1.6 List of *Capsicum baccatum*

	Aji Amarillo: moderately pungent, 10-15 cm long fruit with deep orange when ripe, fruity flavor. Uses: fresh, dried and paste.

Source: Data from Bosland, P., Votava, E.J., *Peppers: Vegetable and Spice* Capsicums, CAB International, Reading, UK, 2012.

Table 1.7 List of *Capsicum pubescens*

	Rocoto: very pungent, very thick wall, shape resembles miniature bell, 5-13 cm long, yellow to red ripe fruit with black seeds. Uses: fresh, dried. and paste. (Photograph courtesy of Bruno B. Defilippi)

Source: Data from Bosland, P., Votava, E.J., *Peppers: Vegetable and Spice* Capsicums, CAB International, Reading, UK, 2012.

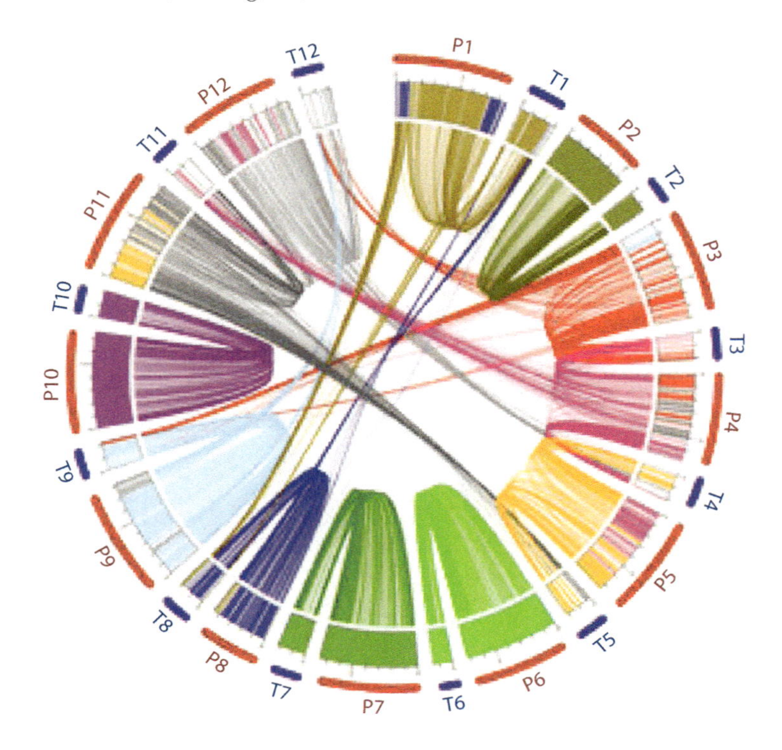

Figure 2.1 *Capsicum* genome, *C. annuum* "CM334," compared with the tomato genome. P1–P12 represent *Capsicum* chromosomes, and T1–T12 represent tomato chromosomes. Lines link the orthologous genes between the two genomes. (Modified from Kim, S. et al., *Nat. Genet.*, 46, 270–278, 2014. With permission.)

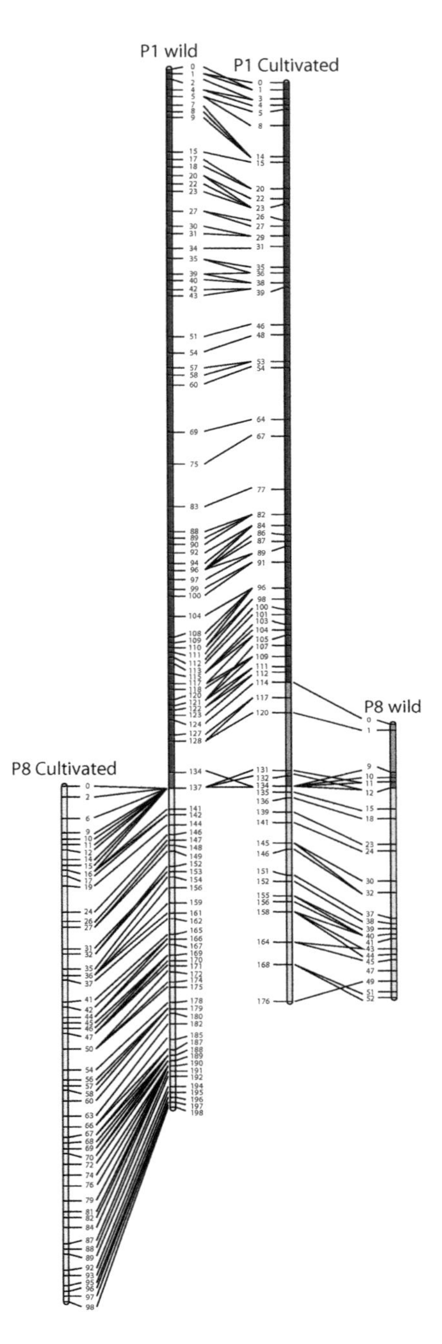

Figure 2.3 *Capsicum* chromosomes show recombination on P1 and P8. Chromosomal regions are colored according to translocation pairs. Non-translocated portions of P1 (P1 wild/P1 cultivated) are in blue, the translocated arms P8 wild/P1 cultivated are in aqua, and P8 wild/P8 cultivated are in yellow. Grey corresponds to pseudolinkage between P1 and P8 wild; red indicates the translocation break point. (Modified from Hill, T. et al., *G3*, 5, 2341–2355, 2015. Figure licensed under the Creative Commons Attribution 4.0.)

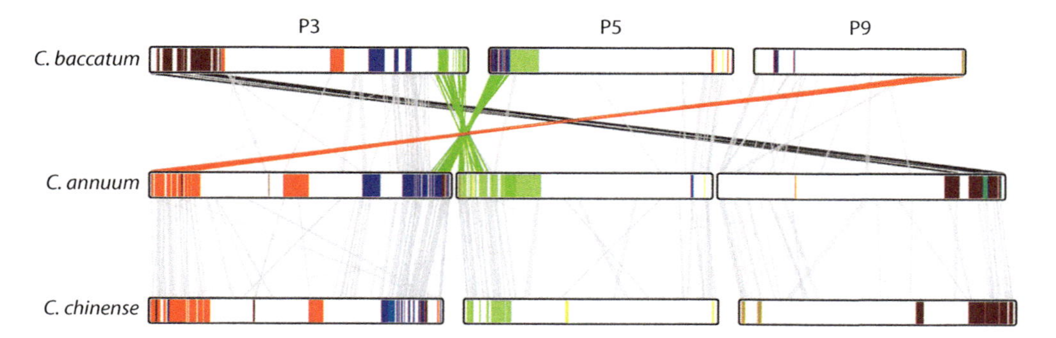

Figure 2.4 A linear comparison of chromosomal rearrangements in three *Capsicum* genomes. Line colors indicate translocations in the ancestral lineage of *C. annuum* and *C. chinense* (red), in *C. baccatum* (green), and in ancestors of *C. annuum* and *C. chinense* or *C. baccatum* (dark grey). (Modified from Kim, S. et al., *Genome Biol.*, 18, 210, 2017. Figure licensed under the Creative Commons Attribution 4.0.)

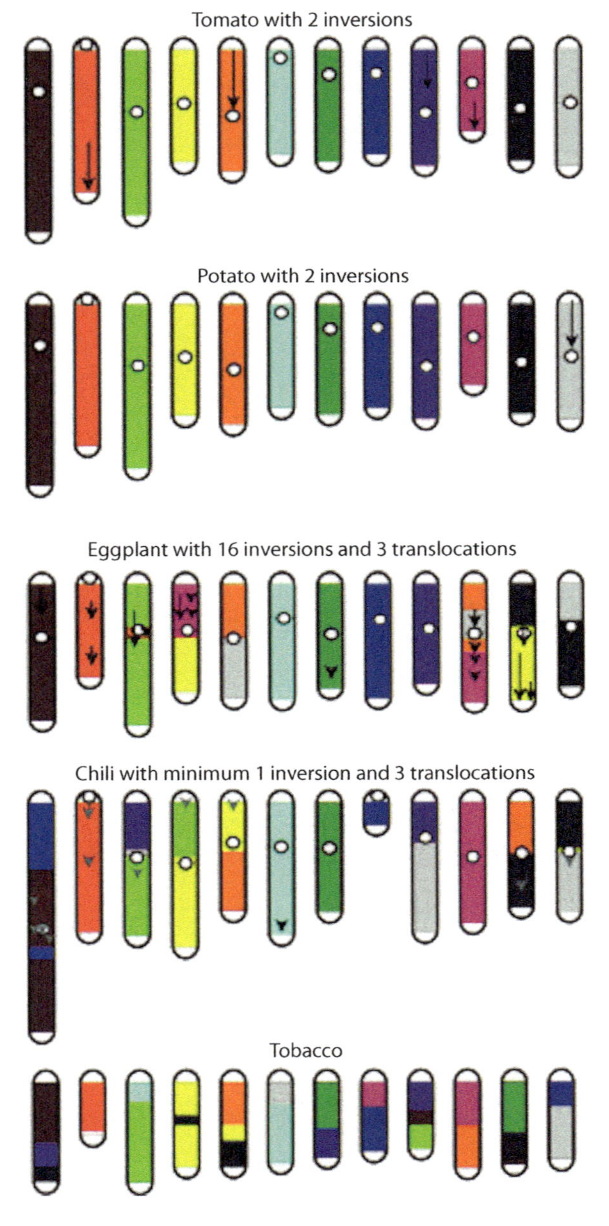

Figure 2.5 Chromosomal evolution in the Solanaceae. Each tomato chromosome is assigned a different color that matches the orthologous counterparts in other species, thus depicting the translocations differentiating these species. Black arrows on the chromosomes indicate inversions; grey arrows represent uncertain inversions. White dots represent putative centromere locations. (Modified from Wu, F., Tanksley, S.D., *BMC Genomics*, 11, 182, 2010. Figure licensed under the Creative Commons Attribution 2.0.)

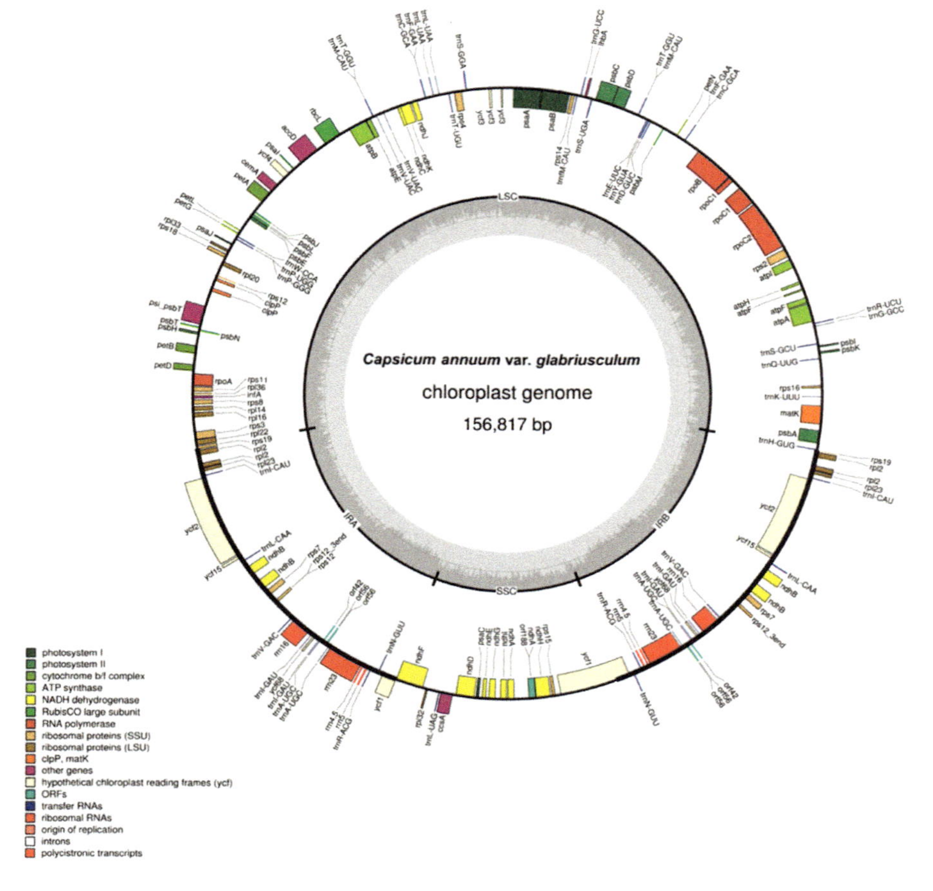

Figure 2.6 Complete chloroplast genome of *Capsicum annuum* var. *glabriusculum*. Inner genes are transcribed clockwise, and outer genes are transcribed counterclockwise. (From Raveendar, S. et al., *Molecules*, 20, 13080–13088, 2015. Figure licensed under the Creative Commons Attribution 4.0.)

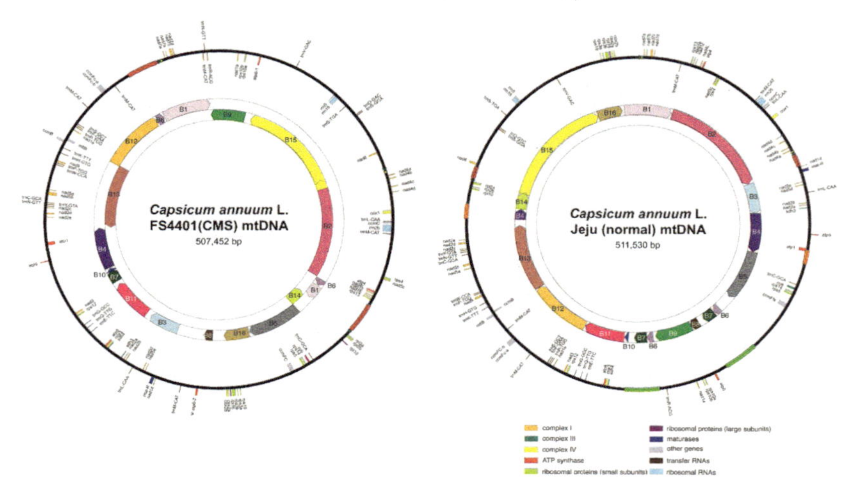

Figure 2.7 Maps of mitochondrial genomes of *Capsicum annuum*, CMS "FS4401" and fertile "Jeju" lines, displaying 16 syntenic sequence blocks (inner circles), B1–B16, between genomes (>95% similarity). Two inner circles are depicted in the CMS "FS4401" to separate blocks with different directions. Different colors denote different functions of the gene products. (Modified from Jo, Y.D. et al., *BMC Genomics*, 15, 561, 2014. Figure licensed under the Creative Commons Attribution 4.0.)

Figure 2.10 Diversity of *Capsicum annuum* complex: flowers and fruits. *C. annuum* var. *annuum* (1–4); *C. annuum* var. *glabriusculum* (5–8); *C. chinense* (9–13), showing the famous Habanero (11) and Bhut Jolokia (12), typical large leaves of the species (13); and *C. frutescens* (14–16) showing typical exerted stigma (14).

Figure 2.11 Diversity: *C. baccatum*, *C. pubescens*, and some wild species. *C. baccatum* var. *pendulum* (1–3), *C. baccatum* var. *baccatum* (4,5), *C. baccatum* var. *praetermissum* (6), *C. pubescens* (7,8), *C. chacoense* (9), *C. eximium* (10), and *C. flexuosum* (11,12).

Figure 3.9 Differential reactions on fruit of *C. baccatum* "PBC80" at mature green (left) and post-mature green, breaking to ripe fruit (right) as inoculated with *Co. scovillei* "MJ8."

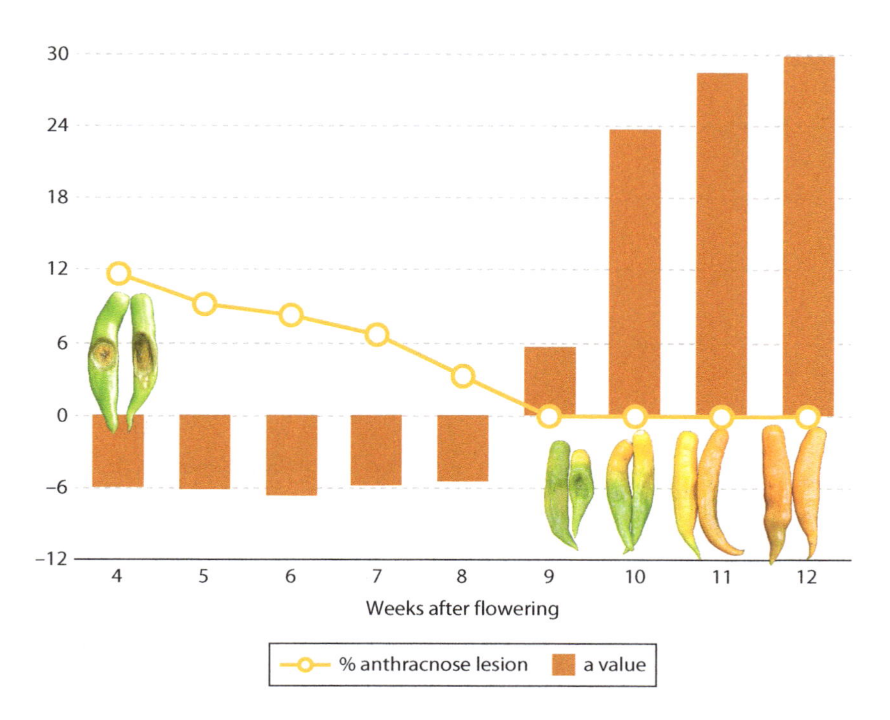

Figure 4.3 Reduced anthracnose symptoms on developing *Capsicum baccatum* "PBC80" fruit. (From Temiyakul, P. et al., *The Journal of King Mongkut's University of Technology North Bangkok* 22, 494–504, 2012.)

Figure 2.8 Hypothetical distribution of the five cultivated *Capsicum* during the discovery of the New World. (Redrawn from Eshbaugh, W.H., *Peppers: Botany, Production and Uses*, CAB International, Wallingford, UK, 2012.)

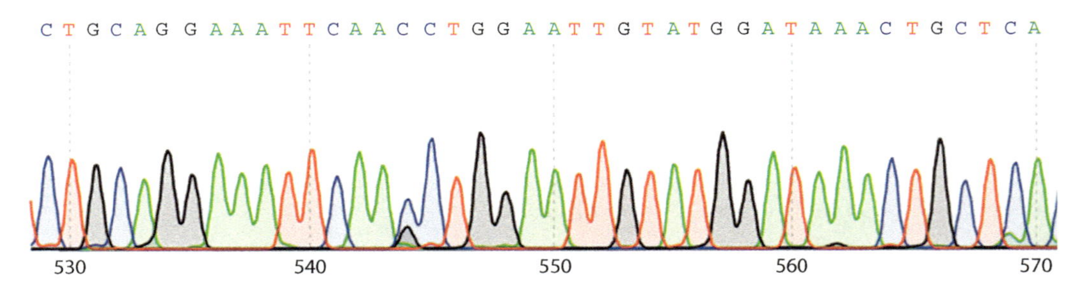

Figure 5.8 A SNP, C/G, is discovered at base 544. (DNA sequence chromatogram courtesy of Dr. Ratri Boonruangrod.)

chapter three

Anthracnose disease in Capsicum

Anthracnose is a major disease of chili, caused by a complex of *Colletotrichum* species. Since 2009, the taxonomy of *Colletotrichum* has dramatically changed due to the application of multi-gene phylogenetic analyses to resolve closely related species. This change has had a significant impact on the nomenclature of the *Colletotrichum* species infecting chili, which will be reviewed in this chapter. This chapter also discusses the host–pathogen interactions based on *Colletotrichum* lifestyles and their pathotype grouping based on differential host reactions. An improved understanding of the etiology of the *Colletotrichum* group of pathogens causing chili anthracnose will lead to precision in resistance breeding and, hence, improved management of the disease.

3.1 *Causal agent and typical symptoms of chili anthracnose*

Chili anthracnose is severe during the wet season in the tropics and subtropics around the world; therefore, chili crops in several Asian and tropical American countries are dramatically affected. The main *Colletotrichum* species reported as the cause of chili anthracnose are *Co. truncatum* (syn. *Co. capsici*), *Co. scovillei* (previously identified as *Co. acutatum*), and *Co. siamense* (previously identified as *Co. gloeosporioides*) (Than et al. 2008; Mongkolporn et al. 2010).

Colletotrichum can infect chili plants at all developmental stages, with more severe disease on the fruit. Typical anthracnose symptoms are usually found on both green and ripe chili fruit as sunken necrotic tissues with concentric rings of wet acervuli (Figure 3.1). However, infected ripe fruit is more predominant. Anthracnose is considered a seed-borne disease (Ranathunge et al. 2012), because heavily infected chili fruit transmits the fungus to the seed. Severe seed infection can result in pre- and post-emergence seedling death; however, in general, most of the infected seed can germinate and show no signs of the infection on the seedlings above the ground. Leaf infection has been reported for *Co. truncatum* (Ranathunge et al. 2012), whereby the pathogen enters a quiescent stage soon after infection of the epidermal cells. The infected leaves remain healthy due to the

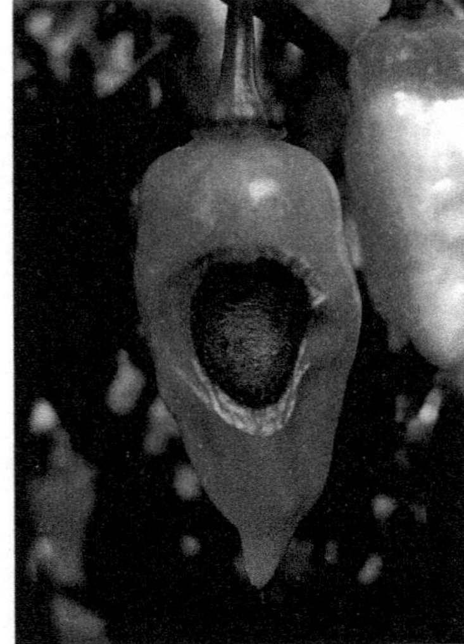

Figure 3.1 Typical anthracnose symptoms on chili fruit in an open field in Thailand. Disease occurs on both ripe (left) and green (right) fruit.

suppressed fungal growth until the leaves senesce and the anthracnose symptoms appear. The infected leaves subsequently detach at senescence; hence, they serve as a primary source of inoculum to reinfect the fruit and leaves. The disease cycle of *Co. truncatum* is illustrated in Figure 3.2.

3.2 Colletotrichum *infection process and lifestyles*

The first stage in the infection process involves conidial attachment to the host surface (Bailey et al. 1992). For *Co. truncatum* on chili fruit, conidial germination started as early as two hours after inoculation (HAI), and by 12 HAI, fully developed dark brown, melanized, globose to irregular-shaped appressoria formed with or without germ tubes (Figure 3.3; Ranathunge et al. 2012). Direct penetration of the cuticle by appressoria began from 12 to 24 HAI, frequently near the cell junctions of the epidermis. Direct hyphal penetration was also observed from the surface mycelium around 72 HAI. The chili fruit cuticle remained intact and was symptomless, although mycelia developed progressively along the cell walls of the sub-epidermal pericarp at four days after inoculation (DAI). At six DAI, the first anthracnose symptoms developed as sunken tissue on the fruit surface. This occurred after dissolution of the sub-epidermal collenchyma cell walls due to mycelial colonization. Despite huge devastation to collenchyma tissue, the fruit cuticle and adjoining epidermal cell layer remained undamaged

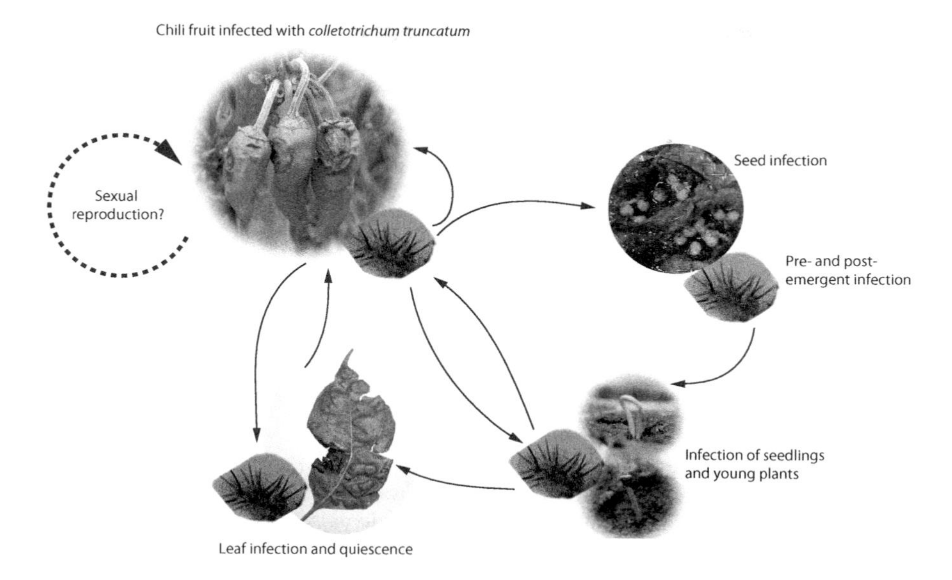

Figure 3.2 Disease cycle of *Colletotrichum truncatum,* according to Ranathunge et al. (2012).

and continued along the surface. The emergence of acervuli on the infected fruit surface began from seven to nine DAI in the prominent dark sunken lesions spreading over the surface. Acervuli originated from a cushion of mycelia formed in epidermal layers of the pericarp close to the cuticle. The epidermis then ruptured and revealed setae, conidiophores, and conidia in tight clusters (Figure 3.3). Figure 3.4 shows some of the morphology of *Colletotrichum* with rod-shaped conidia, *Co. siamense*, and *Co. scovillei*.

Co. truncatum infection hyphae, arising from the appressoria, penetrate the chili fruit surface by direct penetration (Auyong et al. 2015). In some cases, highly branched hyphae, called *dendroid structures*, formed within 24 HAI from appressoria to penetrate the cuticle layer of chili fruit and subsequently into the epidermal cells at 72 HAI (Liao et al. 2012b). The dendroid structures seemed to be specific to the chili fruit and were not found in other parts of chili or other plants. Direct penetration of the cuticle can involve mechanical force or can be promoted by the aid of fungal cutinases. Cutinase hydrolyses plant cuticle to promote fungal penetration and was proved to have a major role in the penetration of *Co. truncatum* through the chili fruit wall (Auyong et al. 2015). Transgenic lines of *Co. truncatum* with reduced expression of cutinase were unable to directly penetrate inoculated chili fruit. However, the ability to infect the epidermal cells was restored in these transgenic lines after the cuticle was mechanically wounded, proving that cutinases are important pathogenicity factors involved in the infection of chili fruit.

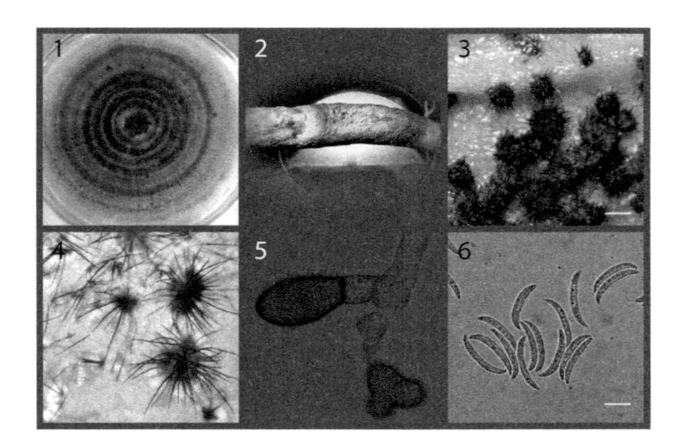

Figure 3.3 *Colletotrichum truncatum*: culture characteristics on potato dextrose agar media (1), culture on chili fruit (2), acervuli on chili fruit surface (3), setae (4), appressoria (5), and falcate shaped conidia (6). Scale bars in photograph 3 = 100 μ, photograph 6 = 10 μm (also applies to photographs 4 and 5). (Photographs 1, 2, 4, and 6 courtesy of Parichat Buaban; photograph 5 courtesy of Paweena Montri.)

Colletotrichum species have complicated lifestyles. To be able to control the disease properly, it is essential to understand the pathogens' lifestyles. Different lifestyles of *Colletotrichum* spp. have been thoroughly reviewed by De Silva et al. (2017b). *Colletotrichum* spp. can have up to four lifestyles: necrotrophic, biotrophic, latent or quiescent, and endophytic. Commonly,

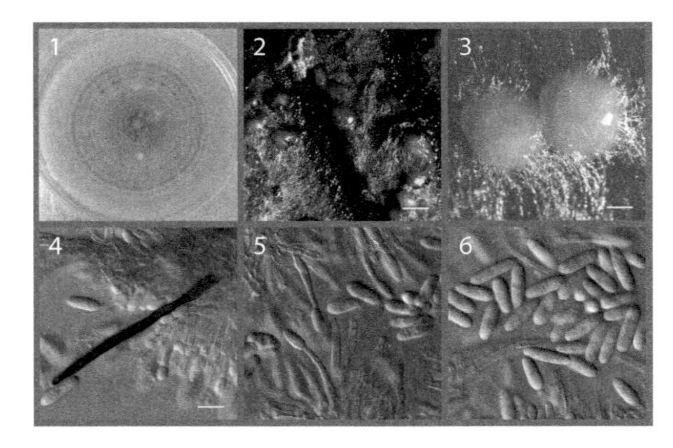

Figure 3.4 *Colletotrichum scovillei* culture with salmon mass conidia on potato dextrose agar (PDA) media (1), *Co. siamense* (2–6): conidiomata on PDA media (2,3), seta (4), conidiophores and conidia (5), oblong shaped conidia (6). Scale bars in photograph 2 = 100 μm, photograph 3 = 200 μm, and photograph 4 = 10 μm (also applies to photographs 5 and 6). (Photograph 1 courtesy of Parichat Buaban; photographs 2–6 courtesy of Dr. Dilani de Silva.)

the chili anthracnose *Colletotrichum* is hemibiotrophic (Kim et al. 2004; O'Connell et al. 2012; Ranathunge et al. 2012); its lifestyle is not truly biotrophic but a mixture of necrotrophy, biotrophy, latency, and quiescence, possibly including a short endophytic stage (Auyong et al. 2012).

The hemibiotrophic *Colletotrichum* generally behaves as a biotroph in the early stages of its life cycle and then switches to a necrotroph in the later stages. A biotroph lives in the host's living cells to feed on the plant nutrients without killing the host cells, while a necrotroph infects, colonizes, and kills the plant cells. Different *Colletotrichum* species have various degrees of hemibiotrophy due to their lifestyle patterns, which are highly regulated by specific gene families (Gan et al. 2016). A rare case of endophytic phase was reported in *Co. truncatum* using a *green fluorescent protein* (*gfp*) transformation (Auyong et al. 2012). The *gfp-Co. truncatum* transformant's growth revealed that the fungus colonized intramurally within the parenchyma tissue of healthy chili fruit without further development of the secondary biotrophic structures. The incidence indicated that the fungus underwent a short endophytic stage before becoming a necrotroph.

Colletotrichum spp. that cause anthracnose may also have a latent or quiescent lifestyle. The *Co. truncatum* life cycle was elucidated on the chili host (Ranathunge et al. 2012). A disease-quiescent stage occurred following leaf infection during vegetative plant growth and served as a potential primary inoculum source for fruit infection. The fungal spore germination and appressoria formation were observed *in planta* as a normal infection process; however, the fungal growth was suppressed soon after the penetration. The pathogen remained quiescent until the chili fruit ripened to continue and complete its life cycle.

3.3 *Taxonomy and diversity of* Colletotrichum *infecting chili*

The genus *Colletotrichum* (Sordariomycetes, Ascomycota) infects over 3,000 plant species globally from herbaceous to woody crops (O'Connell et al. 2012). *Colletotrichum* was first reported in 1790 as *Vermicularia*, and the name *Colletotrichum* was introduced in 1831 (Hyde et al. 2009b). The ability to cause latent or quiescent infections suggests that *Colletotrichum* is one of the most important post-harvest pathogens (Sutton 1992). Recently, *Colletotrichum* was ranked the eighth most important pathogenic genus of fungi in the world (Dean et al. 2012).

The identification of *Colletotrichum* species in the past faced problems mainly due to the uncritical use of species names based on wrong assumptions about host specificity (Cannon et al. 2012). All species were wrongly assumed to be host specific, which led to a large number of described taxa. Species misidentifications cause complications in (1) understanding host–pathogen relationships, (2) developing effective control strategies, (3) establishing cost-effective quarantine programs (TeBeest et al. 1997), and (4) breeding for disease resistance. Conventionally, species identification has relied on conidial and appressorial morphology; the presence of setae, sclerotia, acervuli, and teleomorph state; and mycelial cultural characteristics. However, morphological characters are limited, and large variations of a morphological character occur within species.

The first molecular application to distinguish the *Colletotrichum* species was based on sequence comparison of the internal transcribed spacer (ITS) region of the ribosomal DNA (rDNA) reported in 1992 (Mills et al. 1992; Sreenivasaprasad et al. 1992). Since then, sequences of multiple genes have been used to develop multi-gene phylogenetic analyses for the revision of the taxonomy of *Colletotrichum* species. A polyphasic approach including multi-gene phylogenetic analyses and culture characteristics is now used as the basis to describe species of *Colletotrichum*. The genes currently used in *Colletotrichum* species identification vary by different researchers and different species complexes, since the work has been conducted simultaneously with no agreement among mycologists (Hyde et al., 2009a; Cannon et al., 2012). Genes summarized by Marin-Felix et al. (2017) included ITS, GAPDH (glyceraldehyde-3-phosphate dehydrogenase), CHS-1 (chitin synthase 1), ACT (actin-like protein), HIS3 (histone 3), TUB2 (β-tubulin), GS (glutamine synthetase), CAL (calmodulin), SOD2 (manganese superoxide dismutase), APN2 (DNA lyase 2), and MAT1/APN2 (*ApMat*; mating type gene/DNA lyase 2). Currently there are over 200 *Colletotrichum* species in 11 species complexes (Marin-Felix et al. 2017) with 24 species reported as causal pathogens of chili anthracnose, of which seven belong to the acutatum species complex and nine to the gloeosporioides complex (Mongkolporn & Taylor 2018).

De Silva et al. (2017a) reported that the major *Colletotrichum* species causing chili anthracnose in Australia belonged to the species complexes acutatum, gloeosporioides, and truncatum. The study also identified a new species, *Co. cairnsense*, grouped within the acutatum complex, which also contained the chili anthracnose pathogens *Co. scovillei* and *Co. simmondsii*. The chili anthracnose pathogens in the gloeosporioides complex comprised *Co. siamense* and *Co. queenslandicum*. The three key species of chili anthracnose in Thailand, which were originally identified as *Co. capsici*, *Co. gloeosporioides*, and *Co. acutatum* (Montri et al. 2009; Mongkolporn et al. 2010), were re-classified as *Co. truncatum* (Damm et al. 2009; Ranathunge et al. 2012), *Co. siamense*, and *Co. scovillei* (Damm et al. 2012; De Silva et al. 2017a), respectively. Morphological characters of these three main *Colletotrichum* species that cause chili anthracnose are displayed in Table 3.1.

A recent study from China (Diao et al. 2017) identified 14 *Colletotrichum* species from chili anthracnose, with *Co. fioriniae*, *Co. fructicola*, *Co. gloeosporioides*, *Co. scovillei*, and *Co. truncatum* being the most prominent and *Co.*

conoides, *Co. grossum*, and *Co. liaoningense* newly described. Table 3.2 summarizes the recently identified *Colletotrichum* species causing chili anthracnose from various countries based on pathogenicity and multi-gene phylogenetic analyses. A total of 24 species have been identified (Mongkolporn & Taylor 2018).

Figure 3.5 exhibits a phylogenetic tree, constructed from ITS and TUB2 sequence analysis, of the 24 *Colletotrichum* species that have been reported as chili anthracnose causal pathogens (Mongkolporn & Taylor 2018). The most common species reported are *Co. truncatum*, *Co. siamense*, and *Co. scovillei*. Further studies by De Silva and Taylor have shown that these three species are widely occurring throughout chili-producing areas in Southeast Asia (unpublished data).

3.4 Colletotrichum *pathotypes of chili anthracnose*

Pathogenic variability is a quantitative characteristic used to measure disease severity. Pathogenicity can be measured qualitatively or quantitatively based on the degree of infection of differential host genotypes. Isolates of a species are referred to as *pathotypes* when a subclass or group of isolates can be differentiated from others of the same species by the level of virulence on a specific host genotype (Taylor and Ford 2007).

Montri et al. (2009) and Mongkolporn et al. (2010) identified the existence of *Colletotrichum* pathotypes based on the qualitative differences in an infection of a set of chili species and genotypes. A set of differential chili genotypes with disease scores ranging from 0 to 9 was used to identify pathotypes within three *Colletotrichum* species. The differentiation of host reactions was based on qualitative differences of whether the host was infected (scores 1–9) or not infected (score 0).

Differential host reactions can also be affected by specific host–pathogen relationships. Within the *Co. truncatum* population, differential host reactions were found in the *C. chinense* genotypes, while within *Co. scovillei*, pathotypes were differentiated in the *C. baccatum* genotypes. In addition, fruit maturity has been shown to have an important role in differential host reactions (Temiyakul et al. 2012), with pathotypes being discriminated against by the different stages of fruit maturity (Table 3.3). Three pathotypes were identified within *Co. truncatum* based on the host reactions of ripe fruit and two on mature green fruit. Likewise, one *Co. scovillei* pathotype was differentiated from ripe fruit reactions and three from the mature green fruit reactions (Table 3.4).

In contrast, quantitative differences or levels of aggressiveness based on degrees of infection ranging from low to high on a set of differential chili genotypes were used by Park et al. (2009) to record seven

Table 3.1 Morphological Characteristics and Conidial Measurements of Three Common *Colletotrichum* Species Causing Chili Anthracnose

Taxon	Colony Characteristics	Conidia			Growth Rate (mm/day)
		Length (µm)	Width (µm)	Shape	
Co. truncatum[a]	Flat with entire margin, no aerial mycelium, surface buff, covered with olivaceous-grey to iron-grey acervuli, reverse buff to pale olivaceous-grey, conidia in mass whitish, buff to pale saffron	15.0–20.0	3.5–4.5	Falcate	2.8
Co. scovillei[b]	Flat with entire margin, surface covered with short floccose whitish to pale olivaceous-grey aerial mycelium, margin rosy buff, reverse rosy buff, olivaceous-grey in the center, conidia in mass salmon	14.4–15.0	3.5–4.1	Cylindrical to clavate with one end round, one end ± acute	3.3–3.5
Co. siamense[c]	Cottony, dense greyish white aerial mycelium, pale yellowish to pinkish colony	13.6–15.2	4.8–5.0	Cylindrical	6.9–8.0

[a] Data from Damm et al. (2009), conidia characters based on cultures on synthetic nutrient-poor agar and colony characters based on cultures on oatmeal agar.
[b] Data from Damm et al. (2012), conidia characters based on cultures on synthetic nutrient-poor agar and colony characters based on cultures on oatmeal agar.
[c] Data from Weir et al. (2012), conidia characters based on cultures on potato dextrose agar.

Table 3.2 *Colletotrichum* Species that Cause Anthracnose of Chili Identified Based on Multi-Gene Phylogenetic Analyses, Countries Reported, and Pathogenicity Based on Pre-Wounding (PW) or Non-Wounding (NW) of Fruit

Major Clades	Species	Reported Countries	Pathogenicity	
			PW	NW
Acutatum	*Co. acutatum*	Sri Lanka[a]	?	?
	Co. brisbanense	Australia[a]	?	?
	Co. cairnsense	Australia[b]	Yes	Yes
	Co. fioriniae	China[c]	Yes	?
	Co. nymphaeae	Malaysia[d], India[e], Indonesia[a]	Yes	?
	Co. scovillei	Brazil[f,g,h], China[c,i,j], Indonesia[a], Japan[k], Korea[l], Taiwan[m], Thailand[a,n]	Yes	Yes
	Co. simmondsii	Australia[b]	Yes	No
Boninense	*Co. karstii*	China[c], India[e,o]	Yes	?
Gloeosporioides	*Co. conoides*	China[c]	Yes	?
	Co. fructicola	China[c,i], India[e,p]	Yes	?
	Co. gloeosporioides	China[b,c], India[q]	Yes	?
	Co. grossum	China[c]	Yes	?
	Co. kahawae	India[e]	?	?
	Co. queenslandicum	Australia[b]	Yes	No
	Co. siamense	Australia[b], Brazil[f,h], China[c,i], India[p], Thailand[n]	Yes	Yes
	Co. tropicale	Brazil[f]	Yes	?
	Co. viniferum	China[c]	Yes	?
Spaethianum	*Co. incanum*	China[c]	Yes	?
Truncatum	*Co. truncatum*	Brazil[f], China[c,i], India[e], Pakistan[r], Thailand[n,s,q]	Yes	Yes
—	*Co. brevisporum*	Brazil[f,t,h], China[c,i]	Yes	?

(*Continued*)

Table 3.2 (Continued) *Colletotrichum* Species that Cause Anthracnose of Chili Identified Based on Multi-Gene Phylogenetic Analyses, Countries Reported, and Pathogenicity Based on Pre-Wounding (PW) or Non-Wounding (NW) of Fruit

Major Clades	Species	Reported Countries	Pathogenicity	
			PW	NW
—	*Co. cliviae*	China[c], India[u]	Yes	?
—	*Co. coccodes*	India[e], Serbia[v]	?	?
—	*Co. nigrum*	Argentina[v]	?	?
—	*Co. liaoningense*	China[c]	Yes	?

Source: Reformatted: Mongkolporn, O., Taylor, P.W.J., *Plant Pathology*, 67, in press, 2018.

Pathogenicity tests on fresh chili fruit: "Yes" = successful infection; "No" = unsuccessful infection; "?" = the pathogenicity test not performed.

[a] Damm et al. (2012)
[b] De Silva et al. (2017a)
[c] Diao et al. (2017)
[d] Nasehi et al. (2016)
[e] Katoch et al. (2017)
[f] Silva et al. (2017)
[g] Caires et al. (2014)
[h] De Oliveira et al. (2017)
[i] Liu et al. (2016)
[j] Zhao et al. (2016)
[k] Kanto et al. (2014)
[l] Oo et al. (2017)
[m] Liao et al. (2012a)
[n] Mongkolporn et al. (2010)
[o] Saini et al. (2016)
[p] Sharma & Shenoy (2014)
[q] Montri et al. (2009)
[r] Tariq et al. (2017)
[s] Than et al. (2008)
[t] De Almeida et al. (2017)
[u] Saini et al. (2017)
[v] Liu et al. (2013).

"pathotypes." Sharma et al. (2005), however, claimed that 15 "pathotypes" existed based on quantitative lesion sizes, which were then arbitrarily divided into two reactions: "resistant" and "susceptible." Quantitative disease severity reflects a natural distribution of aggressiveness within a population, ranging from low to high, which, according to Taylor and Ford (2007), is not a true measure of pathogenic differences between isolates. Both studies (Park et al. 2009; Sharma et al. 2005) used a set of differential chili genotypes that expressed different sizes of lesions.

The pathotype identification based on the qualitative differences on chili fruit according to Montri et al. (2009) and Mongkolporn et al. (2010) was found to be in concordance with the genetic studies of anthracnose resistance in *C. chinense* and *C. baccatum*. Genetic analyses of the resistance in chili populations derived from *C. chinense* and *C. baccatum* revealed similar results, in that the resistance was classified based on symptomless fruit; this was an outcome of hypersensitive reaction of the resistance mechanism reported in these two *Capsicum* species (Mahasuk et al. 2009a,b). The genetic analyses in all chili populations also indicated that the resistance at different fruit maturity stages was controlled by different genes. Consequently, pathogenicity testing on different fruit maturities became a basic requirement in the genetic study of chili anthracnose.

3.5 Anthracnose assessment for breeding purposes

Phenotyping is a procedure aiming to evaluate a trait of interest that segregates within a population. A phenotype is an outcome of genotype influenced by environment. The correct phenotypes will help select the desired genotype to be incorporated into an elite cultivar. Phenotyping is consequently the key to the success of breeding programs as well as the localization of resistance genes or other genes of interest.

3.5.1 Field trials and detached fruit bioassay

The assessment of disease resistance is concerned with the interaction of a susceptible plant host, a virulent pathogen, and an environment favorable for disease development (Francl 2001; Scholthof 2007). This interaction is called a *disease triangle*, a plant pathology paradigm. For anthracnose disease in chili, resistance is assessed on the fruit either *in planta* or detached. The *in planta* approach can be conducted in the natural field or a controlled environment. Field assessment is usually set up for varietal trials or germplasm screening, with the trials needing to be grown in environmental conditions favorable for anthracnose development. Known susceptible chili genotypes are grown in a border and in between rows to act as a source for natural inoculum. Alternatively, the disease assessment can be conducted in a closed glasshouse or plastic house or environmental

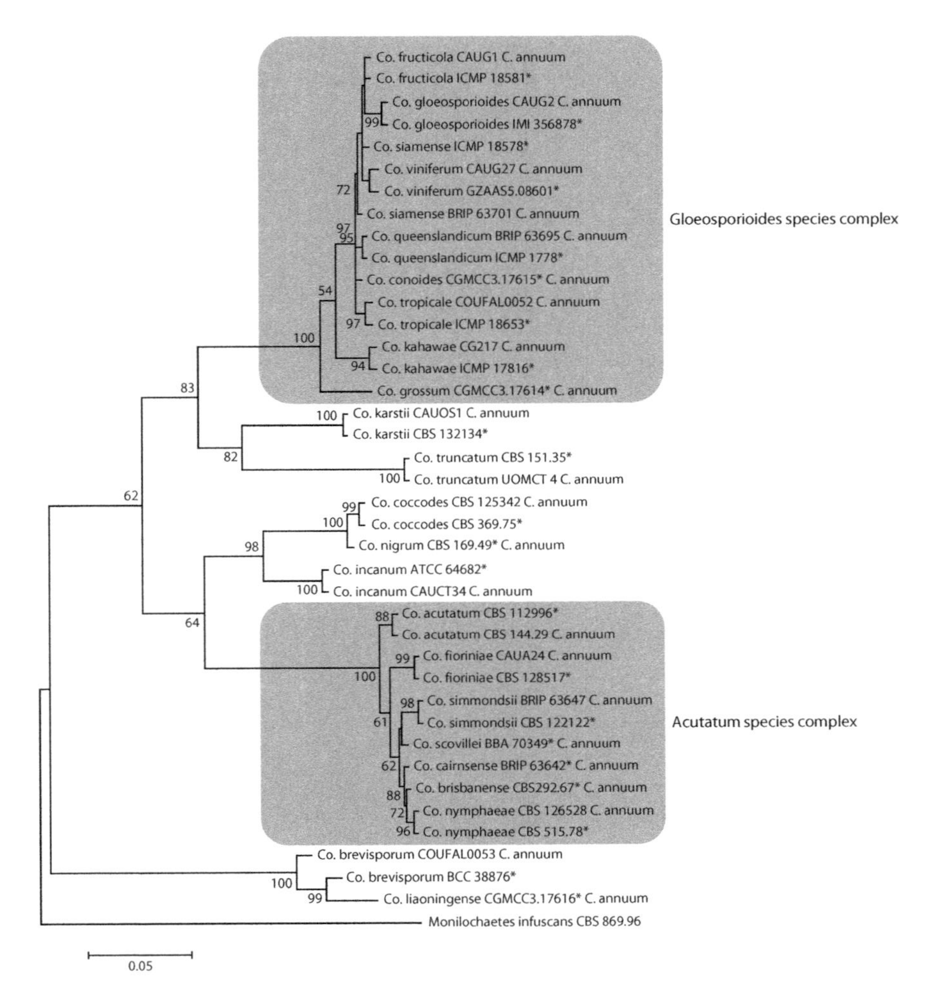

Figure 3.5 Maximum likelihood consensus tree of the combined ITS and TUB2 sequence data of the currently accepted species of 24 *Colletotrichum* that cause anthracnose on *Capsicum* spp. Bootstrap support values >50% are indicated at the nodes and branches. The scale bar shows the number of substitutions per nucleotide position. The tree is rooted with *Monilochaetes infuscans* CBS 869.96. Complex clades follow De Silva et al. (2017a) and Marin-Felix et al. (2017). (Modified from Mongkolporn, O., Taylor, P.W.J., *Plant Pathology, 67*, in press, 2018.)

chamber to maintain high relative humidity as a favorable factor for the pathogen. Spray inoculation is often applied to the fruit or the whole plant. Field trials are subject to the inoculum originating from the local natural *Colletotrichum* species present in the area; however, there may be more than

Table 3.3 Pathotype Identification of *Colletotrichum truncatum* Based on Qualitatively Differential Host Reactions on Mature Green and Ripe Fruit

Pathotype		*Capsicum annuum*		*Capsicum baccatum*		*Capsicum chinense*				*Capsicum frutescens*	
		three genotypes		three genotypes		PBC932		C04714		two genotypes	
ripe	green	ripe	Green	ripe	green	ripe	green	ripe	green	ripe	green
I	I	Y	Y	N	N	Y	N	Y	Y	Y	Y
II	II	Y	Y	N	N	Y	N	N	N	Y	Y
III	–	Y	–	N	–	N	–	N	–	Y	–

Source: Modified from Montri, P. et al., *Plant Disease*, 93, 17–20, 2009; Mongkolporn, O. et al., *Plant Disease*, 94, 306–310, 2010.

Y, infected; N, not infected; –, not available.

Table 3.4 Pathotype Identification of *Colletotrichum scovillei* Based on Qualitatively Differential Host Reactions on Mature Green and Ripe Fruit

Pathotype		*Capsicum annuum, chinense, frutescens* (all genotypes)		*Capsicum baccatum*					
				PBC80		PBC81		PBC1422	
Ripe	Green	Ripe	Green	Ripe	Green	Ripe	Green	Ripe	Green
I	I	Y	Y	N	Y	Y	Y	Y	Y
–	II	–	Y	–	N	–	Y	–	Y
–	III	–	Y	–	N	–	N	–	Y

Source: Modified from Mongkolporn, O. et al., *Plant Disease*, 94, 306–310, 2010.

Y, infected; N, not infected; –, not available.

one species present in equal or, more often, unequal proportions. Infection by multiple species will affect the screening of resistance when breeders wish to select for single gene resistance, especially to specific fungal pathogens.

Highly resistant chili genotypes exist naturally, although these are rarely found in the most popular species of *Capsicum annuum*. Attempts have been made to transfer the resistance from related *Capsicum* species into commercial *C. annuum* cultivars (Mongkolporn & Taylor 2011). Very strong resistance was first identified in some genotypes of the two *Capsicum* species *C. baccatum* and *C. chinense* in 1998 by the World Vegetable Center of Taiwan. These resistant chili genotypes have been extensively incorporated into chili breeding programs in Asia (Pakdeevaraporn et al. 2005; Yoon & Park 2005; Yoon et al. 2006; Lin et al. 2007; Kim et al. 2008a,b; Mahasuk et al. 2009a,b; Sun et al. 2015; Suwor et al. 2015). It is more favorable to conduct the disease assessment in controlled conditions, such as in a laboratory or a glasshouse, for mass screening of the resistance. In such controlled environments, all the factors influencing the disease severity, including pathogen virulence and inoculum quantity, and favorable conditions for infection are consistently controlled. Only one factor, chili genotype, is left to be investigated. The goal of the fruit bioassay is to find true genetic resistance, and to achieve this goal, the inoculation procedure is the key factor in the process. Inoculation with a prescribed spore inoculum concentration ensures the delivery of a constant amount of virulent fungal spores to the host tissues. For chili anthracnose, there are two basic approaches widely used, whereby the fruit is either wounded or not wounded prior to the inoculations.

The purpose of wound inoculation is to ensure direct entry of the germinated spores into the host cells, which bypasses the primary barrier of defense—the cuticle on the chili fruit. The wound on the chili fruit is usually around 0.5–1.0 mm in both depth and diameter. This can be made by a pinprick or a special microinjection syringe (Figure 3.6; Kanchana-udomkan et al. 2004). The first symptom of anthracnose is a water soak lesion that develops at the inoculation site on the third day after inoculation (Montri et al. 2009).

The non-wound inoculation delivers the fungal spores onto the chili fruit by either dropping or spraying. The dropping method is usually conducted on detached chili fruit in a laboratory (Kanchana-udomkan et al. 2004). The spraying method is flexible in that the inoculation can occur either directly on the plant or on the detached fruit. Without wounding, anthracnose symptoms take longer to develop, because the infection hyphae of germinated fungal spores have to penetrate through the cuticle barrier on the fruit pericarp. Several chili genotypes of all *Capsicum* species inoculated using the non-wound inoculation method did not become

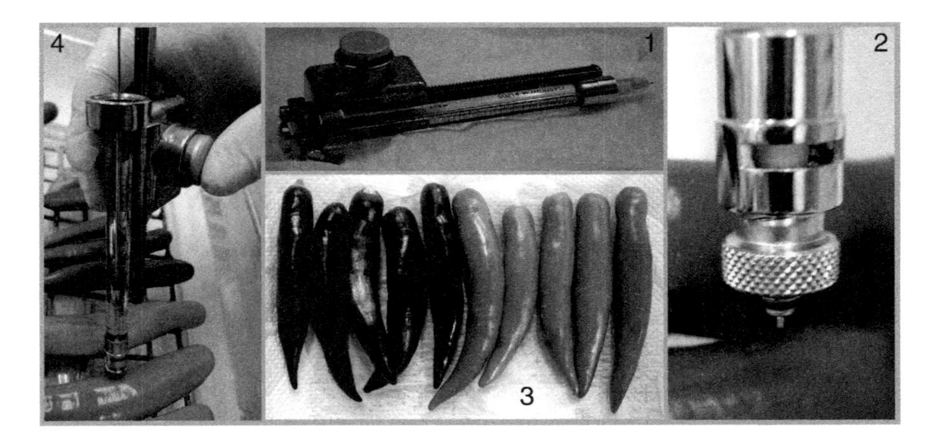

Figure 3.6 A microinjection inoculation method: (1) a gas-tight microsyringe model 1705 TLL and a dispenser PB600-1 (Hamilton, Switzerland) attached to a cost-effective modified needle of medical hypodermic type (needle tip cut to 1 mm long), (2) a modified 1-mm needle from Hamilton, (3) surface sterilized mature green and ripe fruit prepared for inoculation, (4) fruit inoculation by injection.

infected (Kanchana-udomkan et al. 2004; De Silva et al. 2017a), indicating that the cuticle played a role in resistance in these chili genotypes.

A high-pressure spray inoculation method has also been developed (Figure 3.7; Mahasuk et al. 2013), which was thought not to cause a wound on the chili fruit. The high-pressure spray delivered the fungal inoculum onto chili fruit using an airbrush and an air compressor. However, similar results were obtained from both microinjection and high-pressure spray methods with a delayed infection by the latter. The chili fruit surface was most likely slightly wounded, thus breaking the defense of the cuticle. For breeding purposes, either microinjection or high-pressure spray can be used. However, microinjection is more advantageous than the high-pressure spray in that it requires less inoculum and fruit preparation (Mahasuk et al. 2013). In addition, a double inoculation can be performed on a chili fruit (Temiyakul et al. 2010) with microinjection, whereby the fruit is injected twice on the opposite sides with two different pathogen isolates. As a result, the double inoculation helps save resources and time. Figure 3.8 displays fruit incubation after being inoculated.

The wound method has been more widely used than non-wound inoculation due to the more rapid disease development. The speed of disease development is important for the chili anthracnose bioassay on detached fruit, so that the disease evaluation can be completed before the fruit senesces. However, the wounding inoculation is extremely efficient; therefore, only a few chili genotypes of *C. baccatum* and *C. chinense* are

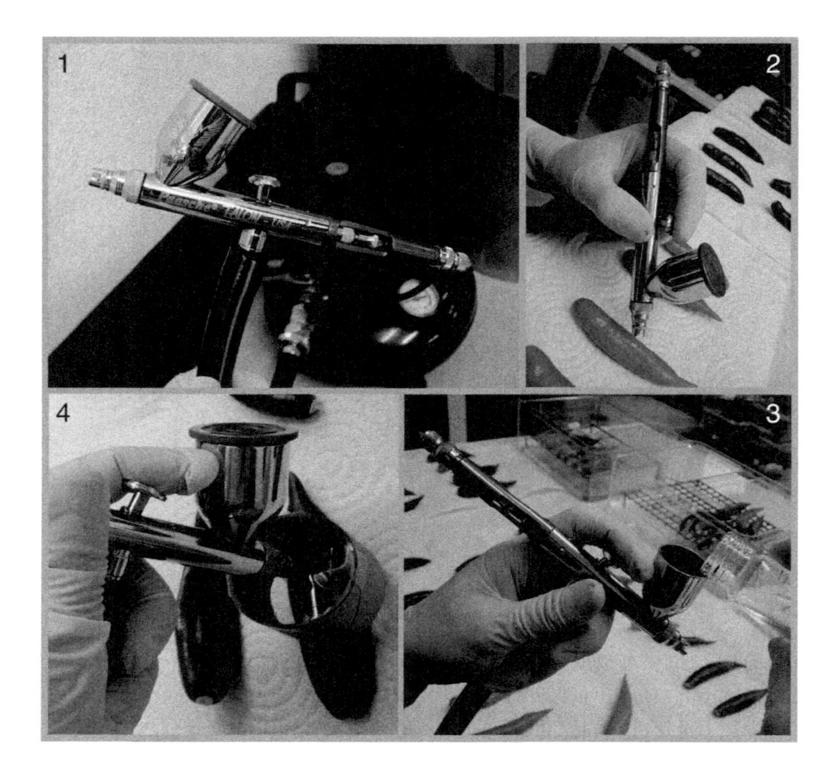

Figure 3.7 A high pressure sprayer: (1) an airbrush model TG-3F (Paasche®, USA) and an air compressor model ACA 201 (K. Setthakit, Thailand); (2) the pressure adjusted to 2 kg cm^{-2} to deliver approximately 6,000 spores; (3) filling up prepared inoculum; (4) the spray position on the fruit and the sprayer height above the fruit is controlled by a tube. (Photographs courtesy of Dr. Pitchayapa Mahasuk.)

not infected (Mongkolporn et al. 2010). On the other hand, for pathogenicity studies to identify *Colletotrichum* species, the wounding fruit bioassay is also favorable, although few studies have used both wound and non-wound methods. *Colletotrichum*, as a natural pathogen, successfully infects host plants without wounding. Without the non-wounding inoculation included in the pathogenicity study, a *Colletotrichum* isolate may not be confirmed as a true pathogen to chili. Therefore, a minimum protocol for pathogenicity testing on non-wounded fruit is recommended.

3.5.2 Anthracnose resistance assessment

Disease severity reflects the quantity of pathogenic infection, which is important in deciding whether the plants are resistant or susceptible. This knowledge of disease severity is useful for evaluating germplasm,

Figure 3.8 Post-inoculation fruit incubation: (1) fruit incubation in a humidity chamber, (2) in a plastic clear box, and (3) disease evaluation on differential chili genotypes in the pathotype study.

breeding programs, and disease management solutions. Disease severity on plants is basically determined by measuring or estimating the size of the lesions as disease symptoms develop on host plants. Disease measurements need to be reliable, precise, and accurate (Bock et al. 2010). Broadly, methods for disease assessment are divided into two types: visual estimation, using disease rating scales, and measurement using image technology. Traditionally, visual estimation is conducted by trained raters. However, Bock et al. (2008) proved that visual disease rating was less precise than using image analysis software. Substantial variation was observed within an individual rater when performing different assessments and between raters. Image-based disease assessment is consequently becoming popular (Bock et al. 2010; Mutka & Bart 2015; Rahaman et al. 2015). The benefits of image-based assessment are the reduction in human error when evaluating a large population and the technology's ability to measure disease symptoms at an early stage that is not yet visible. For example, plants that are stressed in the early stage of infection alter their patterns of chlorophyll fluorescence emission, so these changes can be detected by fluorescence imaging (Mutka & Bart 2015).

At minimum, the image analysis requires visible light and a digital camera to produce images. The images are then analyzed by computer software to differentiate the healthy and diseased areas (Bock et al. 2010). Thermal infrared detects the increased temperature of infected plant tissues (Mutka & Bart 2015). Hyperspectral, or spectroscopy, imaging is the newest method and is widely used in agricultural research, such as monitoring fruit quality, creating land use maps, and detecting insect, weed, and pathogen infestation (Bock et al. 2010). A hyperspectral image is created from an interaction of the subject with three-dimensional (two spatial and one spectral) electromagnetic spectra at every pixel in an image that absorbs and reflects the energy. In addition to disease measurement, the imaging technology is widely used for other phenotypes, such as quality of fruit and grain, water content and flow, pigment composition, root architecture, and photosynthetic performance (Rahaman et al. 2015).

Currently, visual disease assessment is still in use, especially for chili anthracnose resistance assessment. The disease symptoms are quantified into four dimensions: disease intensity (quantity of disease present in a population), prevalence (proportion of places where the disease is detected), incidence (proportion of plants or plant units infected of total number assessed) and severity (area of sampling unit showing symptoms) (Nutter et al. 1991).

Visual disease assessments by raters or measurement devices can be incorrect. The quality of the assessments requires precision, reliability, reproducibility, repeatability, accuracy, and agreement, so that comparisons can be made between assessment data (Bock et al. 2010). In the case of chili anthracnose, various assessment methods are being used by different research groups, yielding different results and outcomes with the same resistance sources and inoculation method. Therefore, the quality of "agreement" in chili anthracnose assessments is being questioned. Assessing fewer, larger lesions is less prone to error than assessing small, random, and uniformly distributed lesions. In some cases, averaging data does not reflect actual disease severity, especially when there is a constant inconsistency of the symptoms among samples. This problem is very common in chili anthracnose, in which the cause of the inconsistent symptoms may have been related to the fruit maturity (Figure 3.9); therefore, solving the problem by increasing replication (fruit number) may possibly result in more error.

Disease incidence can be assessed by field screening or laboratory-based bioassay, especially when the spray inoculation is applied. Within this parameter, two more dimensions can be achieved: disease severity and the area under disease progress curve (AUDPC). The disease severity is determined based on the proportion of fruit infected per plant (Gottwald et al. 1989). AUDPC estimates the disease progress by combining multiple disease parameters at different sampling times into a single value (Simko & Piepho 2012). The AUDPC is calculated from the area under the curve as the disease progresses. The higher the AUDPC, the faster the disease develops.

On the other hand, when the inoculation is performed on detached fruit, lesion size is commonly measured as well as the disease incidence (Voorrips et al. 2004; Kim et al. 2008a,b) and AUDPC (Kanchana-udomkan et al. 2004; Silva et al. 2014). The lesion size is measured by either length, diameter, or area (Park et al. 1990a,b; Lin et al. 2002, 2007; Kim et al. 2010). Some assessments take the fruit size into account to calculate the proportion of lesion size to fruit size and rate this as a disease score (Montri et al. 2009; Silva et al. 2014).

Suggested ways to improve the accuracy of visual disease assessment are by training the raters and using standard area diagrams. In chili anthracnose, Montri et al. (2009) developed sets of disease diagrams for different chili fruit shapes and sizes. A diagram set consisting of six computerized pictorial chili fruit that represent disease severity scores that vary from 0 to 9 is shown in Figure 3.10. Each score was defined as percentage of lesion in proportion to the overall fruit size. Several diagram sets were made according to different chili fruit types, and these diagrams were printed out as large as the actual chili fruit and used for visual assessment by trained raters.

The interpretation of the disease rating scores as related to phenotype is not a standard protocol of grouping or grading. Disease scores segregating within a population distribute from low to high values, which can be phenotypically classed from highly resistant to highly susceptible or, alternatively, from highly susceptible to highly resistant. For chili anthracnose, phenotypic classification varies depending on the genetic combination of the chili parents that generate the segregating population, the *Colletotrichum* isolate, the fruit bioassay, and the resistance assessment

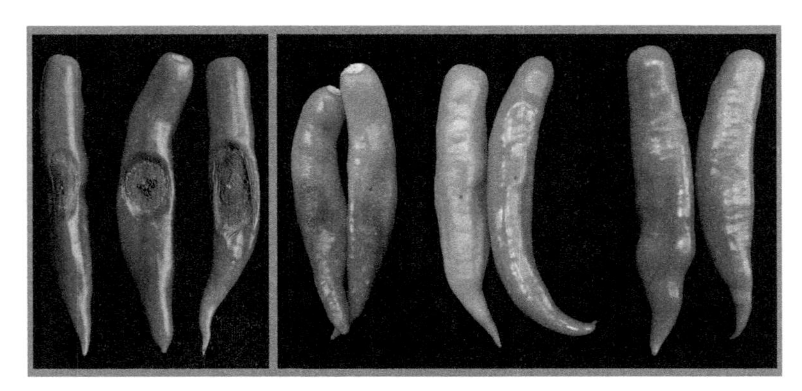

Figure 3.9 (See color insert.) Differential reactions on fruit of *C. baccatum* "PBC80" at mature green (left) and post-mature green, breaking to ripe fruit (right) as inoculated with *Co. scovillei* "MJ8."

method. Therefore, for breeding purposes, the resistant parent determines the *Colletotrichum* isolate and the resistance assessment method to be used in the breeding program. For example, *C. chinense* "PBC932" was susceptible to *Co. scovillei* but resistant to *Co. truncatum* (Montri et al. 2009; Mongkolporn et al. 2010). "PBC932" was used as the resistant donor to improve the resistance to *Co. truncatum* in a *C. annuum* cultivar. The resistance recovered in the F2 progeny was as high as the hypersensitive reaction (HR) derived from the "PBC932" parent (Pakdeevaraporn et al. 2005; Mahasuk et al. 2009a). The significance of phenotyping is that the correct phenotype accommodates the genetic analysis of a disease resistance trait and ultimately, the discovery of the resistance genes via genome mapping

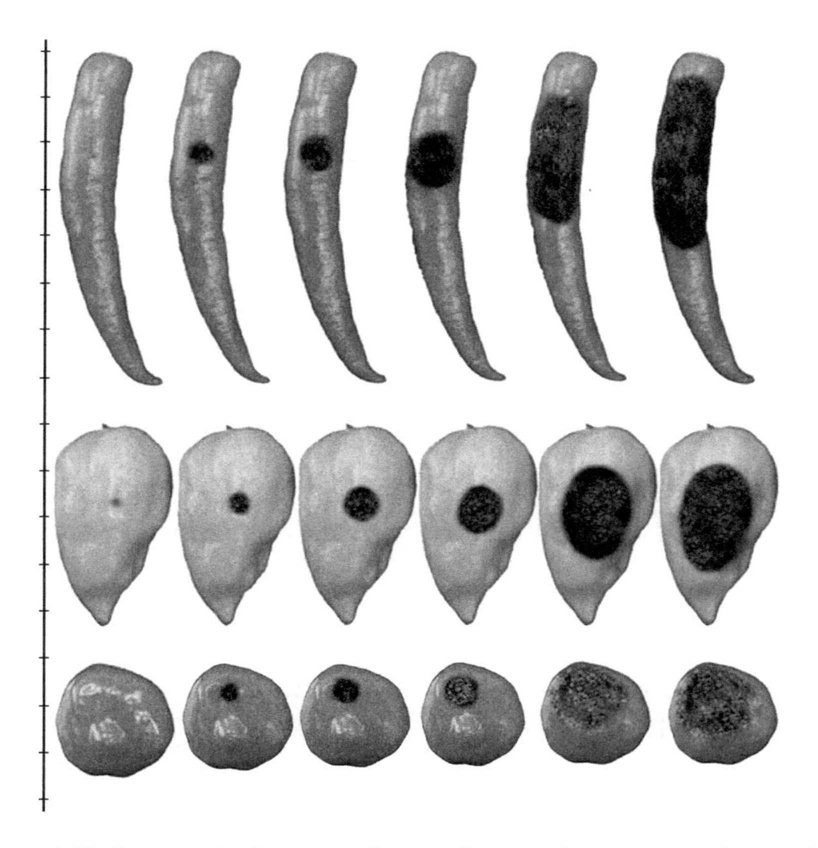

Figure 3.10 Computerized image scales to estimate anthracnose severity ranging from scores 0 to 9 (left to right): based on lesion size proportional to the overall fruit size of three different chili fruit shapes and sizes, scores 0 = HR, 1 = 1–2%, 3 = >2–5%, 5 = >5–15%, 7 = >15–25%, and 9 = >25%. Scale bars on the left represent 1 cm interval. (Modified from Montri, P. et al., *Plant Dis.*, 93, 17–20, 2009.)

technology. Without correct phenotyping, the genes responsible for or DNA markers linked to the trait will not be discovered.

3.5.3 Defense mechanisms

The resistance reaction on resistant chili fruit in response to *Co. truncatum* and *Co. scovillei* is an HR, which has been identified in a few chili genotypes of *C. chinense* "PBC932," *C. baccatum* "PBC80" and "PBC81" (Kim et al. 2004; Montri et al. 2009; Mongkolporn et al. 2010; Ranathunge et al. 2012), and "Bhut Jolokia" cultivar (Mishra et al. 2017). "Bhut Jolokia" is a natural interspecific hybrid of *C. chinense* and *C. frutescens*, according to Bosland & Baral 2007), and is recognized as one of the world's hottest chili cultivars. The HR (Figure 3.11) is a form of programmed cell death involving a thickened fruit cuticle layer (Kim et al. 2004), cytoplasmic shrinkage, chromatin condensation, mitochondrial swelling, vacuolization, and chloroplast disruption (Coll et al. 2011; Stael et al. 2015).

In general, plants have two defense systems: passive or pre-existing and active or induced mechanisms (Kim et al. 2004). Pre-existing defense refers to physical or primary barriers that protect the plants from pathogens, such as thickened cuticle layers, rigid cell walls, cuticular lipids, trichomes, antimicrobial enzymes, and secondary metabolites. In this category, *Capsicum* structures include the cuticle thickness and cuticular lipid composition, which may play a role in primary defense; however such structures have not been reported in anthracnose resistance. Glandular trichomes of different *Capsicum* species were studied for their direct and indirect defenses (van Cleef 2016) and it was suggested that one type of trichome produced a sesquiterpenoid compound that acted against greenhouse spider mites (*Tetranychus urtica*). Phytoalexins are a heterogeneous group of compounds with low–molecular mass secondary metabolites and antimicrobial activity (Ahuja et al. 2012). Capsidiol and capsicannol are phytoalexins identified in anthracnose-infected chili fruit (Adikaram et al. 1982). Both inhibited fungal growth, with the capsidiol having a role in immature fruit and the capsicannol in ripe fruit.

Induced defense is a complex network involving molecular recognition and signaling and is broadly divided into two layers: pathogen-associated molecular patterns triggered immunity (MTI; formerly called *basal* or *horizontal resistance*) and effector triggered immunity (ETI; formerly called *R-gene based* or *vertical resistance*) (Jones & Dangl 2006; Muthamilarasan & Prasad 2013; Stael et al. 2015). After passing the primary defense, the successful pathogens are recognized by a class of transmembrane pattern recognition receptors (PRRs). The successful recognition consequently activates the MTI. In this defense layer, the pathogenic molecules are commonly conserved in the same class of microbes, such as bacterial flagellin, xylanase, ergosterol, and lipopolysaccharides. The first immunity

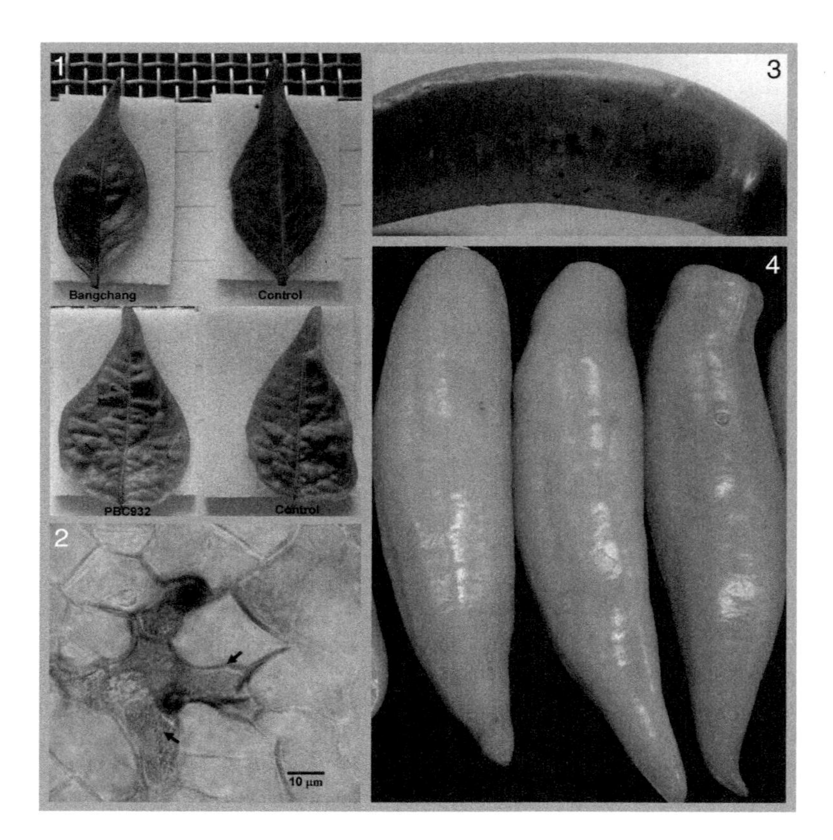

Figure 3.11 Hypersensitive reaction (HR): differential leaf reactions on susceptible "Bangchang" cultivar showing necrosis and resistant "PBC932" showing HR in comparison to healthy leaves (control; 1); a histological image of the HR reaction showing thickened cell walls and cytoplasm aggregation around the infection site (2); HR responses on the "PBC80" fruit inoculated by high-pressure spray (3) and by microinjection (4). (Photographs 1 and 3 are from Mahasuk, P. et al., *Plant Breeding*, 128, 701–706, 2009a; photographs 2 and 4 are from Mahasuk, P. et al., *Breed. Sc.*, 63, 333–338, 2013.)

layer or MTI generally involves a calcium burst, the production of reactive oxygen species (ROS), callose deposition at the cell wall, the activation of mitogen-activated protein kinase (MAPK) cascades, the expression of defense-associated genes, and the production of ethylene. Virulent pathogens produce virulence effectors that can inhibit the MTI, thus resulting in susceptibility of the host.

On the other hand, plants have a second layer of defense, the ETI. A pathogen effector that overcomes the MTI is specifically recognized by intracellular nucleotide binding leucine-rich repeat (NB-LRR) protein, encoded by most *R* genes, and thus activates the ETI inside the plant cells.

ETI is an accelerated and amplified MTI response resulting in disease resistance, commonly an HR at the infection site (Jones & Dangl 2006; Stael et al. 2015). Plants that fail to recognize the pathogen effector become infected.

HR appears to be an integrated outcome of plant defense via both ETI and MTI; the chloroplast plays a central role in HR by housing a very important source of defense signaling molecules, i.e., ROS) reactive nitrogen oxide intermediates (NOI), and defense hormones (salicylic and jasmonic acids) (Coll et al. 2011), with calcium playing a secondary messenger role (Stael et al. 2015). Increased ROS levels not only have a direct toxic effect on the pathogens but also act as signaling molecules to activate the oxidative burst leading to the HR.

The HR has been morphologically identified in anthracnose-resistant chili genotypes, but understanding of its molecular mechanisms is still limited. Several studies have reported various genes involving the HR responses in chili. A recent study of gene expression in relation to HR in "Bhut Jolokia" wound inoculation with *Col. truncatum* revealed 17 defense-related genes involved in multiple signaling pathways, i.e., defense-responsive, pathogenesis-related (*PR*), and transcription factors (Mishra et al. 2017). The defense-responsive genes *plant defensin* (*PDF1.2*), *lipooxygenase* (*Lox3*), *allene oxide synthase* (*AOS*), *ACC synthase* (*ACS2*), and *phenylalanine ammonia lyase* (*PAL3*) were reported to have important roles in the production of defense-related hormones such as jasmonates, salicylates, and ethylenes. For the *PR* group, *PR-2* and *PR-5* were found to be highly active against the pathogen infection. *PR-2*, encoding β-glucanase, and *PR-5* had been reported to act against a wide range of fungal pathogens. The transcription factors have important roles in signaling pathways and usually bind to the defense-responsive genes. Previous studies reported a cytochrome P450 as a defense-responsive gene identified in a different chili genotype of *C. annuum* (Oh et al. 1999a), and *PR-10* was identified in *C. baccatum* "PBC80" (Soh et al. 2012). *PR-10* proteins were reported to be involved in intracellular defense; they were induced by abiotic stresses and capable of cleaving pathogenic RNAs. In addition, an esterase gene (*PepEST*) isolated from an HR in response to chili anthracnose was proved to inhibit *Colletotrichum* appressoria formation (Kim et al. 2001).

3.6 Conclusions and remarks

Chili anthracnose is caused by a complex of *Colletotrichum* species. Based on multi-gene phylogeny, 24 *Colletotrichum* species infecting chili have been identified, with the three main species being *Co. scovillei, Co. truncatum*, and *Co. siamense*, the common causal agents globally. Correct taxonomy of *Colletotrichum* species is helpful in understanding their host–pathogen relationships, developing effective disease control strategies, and establishing cost-effective quarantine programs. *Colletotrichum* species infecting chili are hemibiotrophs, whose lifestyle may involve necrotrophy, biotrophy, latency, or quiescence, and sometimes a short endophytic stage.

The ultimate goal for sustainable control of chili anthracnose is to breed for durable anthracnose resistance, which should contain different resistance gene loci to a broad range of *Colletotrichum* species and pathotypes. Pathotype identification based on qualitative difference of infection on differential chili cultivars with known resistance genes may lead to future *Colletotrichum* race identification. The races of *Colletotrichum* species that cause chili anthracnose have never been identified. A major obstacle to the identification of races in chili anthracnose is the lack of standardization of the resistance assessments (bioassay and the measurement of and determination of resistance) between research groups. Resistant chili genotypes are useful in identifying *Colletotrichum* pathotypes, because they provide differential reactions to different *Colletotrichum* isolates. Breeding for resistance to races would broaden the resistance base of chili cultivars through gene pyramiding of multiple resistant genes into one cultivar. The establishment of a uniform procedure to identify pathotypes or races of *Colletotrichum* species for anthracnose assessment, covering a standard set of chili host genotypes and consolidated inoculation and disease evaluation methods, would be necessary.

chapter four

Breeding for anthracnose resistance in Capsicum

Resistance to chili anthracnose is rare in *Capsicum annuum*, which is the most popular and important *Capsicum* species worldwide. The search for anthracnose resistance has identified hypersensitive reaction (HR) resistance in some genotypes of *C. baccatum* and *C. chinense*. It is therefore necessary to introduce the HR resistance found in *C. baccatum* and *C. chinense* into elite cultivars of *C. annuum* by introgression via wide hybridization. *Capsicum*-wide hybridization may be achievable with or without an aid, depending on the degree of species compatibility. However, the key to successful breeding for anthracnose resistance relies heavily on the selection process for the resistance trait, which is currently based on bioassay of the fruit. In chili, fruit maturity factor is well documented as a key player in the expression of anthracnose resistance.

4.1 Sources of anthracnose resistance in Capsicum

Breeding for resistance to anthracnose in chili started before 1990 by Korean and Indian chili breeders (Park et al. 1990a,b; Ahmed et al. 1991). Their breeding programs mainly relied on resistance obtained from *C. annuum* sources, which were not highly resistant. Thirty-four *C. annuum* and one *C. frutescens* cultivar, collected from Korea, India, Thailand, and the United States, were screened for anthracnose resistance with foliar spray inoculation. None of them showed high resistance (Hong & Hwang 1998). The lack of resistance sources within *C. annuum* is one of the more serious problems that have hampered the success of anthracnose-resistance improvement.

The major sources of resistance to anthracnose that have been widely used in Asia are HR resistance identified in *C. baccatum* and *C. chinense* (AVRDC, 1999). To be able to manage the screening of a large number of chili germplasm, a standard resistance assessment method needs to be performed in controlled conditions. A fast and accurate inoculation method has been developed, which involves the injection of a small volume of inoculum into detached chili fruit in a laboratory with controlled temperature, light, and humidity. The inoculum volume is usually 1 μl delivered by a microsyringe; therefore, the inoculation method is called *microinjection*. Microinjection (Figure 3.5) comprises an airtight microsyringe attached to a small needle, which is specially cut to 0.5–1.0 mm long (Kanchana-udomkan et al. 2004; Montri et al. 2009). Every single 1 μl injection usually delivers approximately 1000–5000 spores. To achieve a fast and accurate injection, a dispenser is attached to the microsyringe for the rapid release of an accurate inoculum volume. Since the microinjector set is expensive, and the needle's desired length is not commercially available, the method has undergone various modifications. Basically, the chili fruit is wounded by pinpricking and then inoculated with a drop of inoculum (Kanchana-udomkan et al. 2004) or a fungal agar disc (Kim et al. 2010; Liu et al. 2016). Microinjection, as well as the modified method, is widely used in chili breeding programs in Asia (Lin et al. 2002; Voorrips et al. 2004; Pakdeevaraporn et el. 2005; Kim et al. 2007b; Lin et al. 2007; Sucheela 2012; Syukur et al. 2013; Suwor et al. 2015; Liu et al. 2016; Zhao et al. 2016), Australia (De Silva et al. 2017), and Brazil (Silva et al. 2014).

The discovery of HR resistance in *C. baccatum* and *C. chinense* by the World Vegetable Center in 1998 has had a great impact on the chili breeding programs in Asia. The resistant *Capsicum* sources have been widely shared and incorporated into elite chili cultivars. Several chili populations have also been developed for genetic studies and mapping the resistance genes and their linked markers. However, chili germplasm screening for anthracnose resistance is continuing, without much success in identifying the resistance in *C. annuum*. Over 300 chili accessions from five cultivated species—*C. annuum, C. baccatum, C. chinense, C. frutescens*, and *C. pubescens*—were screened by the microinjection method (Yoon et al. 2004). Resistance was identified only in *C. baccatum*, although the majority (nearly 300) of the *Capsicum* species screened were *C. annuum*. More recent screening attempted in Brazil was able to identify more resistant genotypes in *C. baccatum* and *C. chinense* (Silva et al. 2014).

The three most popular hypersensitive reaction-resistant genotypes are *C. chinense* "PBC932," *C. baccatum* "PBC80," and "PBC81" (Figure 4.1). The resistance mechanism was characterized as a hypersensitive reaction (HR) Kim et al. 2004; Mahasuk et al. 2009a). HR will be described in Chapter 3. Introgression of the resistance into elite *C. annuum* cultivars is not simple and depends on the crossability between the *Capsicum* species. The *Capsicum* genus forms three complexes: *C. annuum, C. baccatum*, and *C. pubescens*, based on their crossability (Pickersgill 1971, 1997). The *C. annuum* complex comprises *C. annuum, C. chinense, C. frutescens, C. chacoense*, and *C. galapagoense*, while *C. baccatum* is in a different complex. *C. chinense* is more closely related to *C. annuum* than *C. baccatum* is. Therefore, interspecific crosses between *C. annuum* and *C. chinense* have a higher success rate than crosses between *C. annuum* and *C. baccatum* (Martins et al. 2015).

C. chinense "PBC932" was successfully crossed with *C. annuum* cultivars (Pakdeevaraporn et al. 2005; Kim et al. 2007b; Lin et al. 2007;

Sun et al. 2015). These "PBC932"-derived populations have been used to locate the anthracnose resistance genes. Since they are in different *Capsicum* complexes, a successful cross between *C. annuum* and *C. baccatum* is not easy. Their hybrid seed is not viable, mostly due to embryo abortion. An anatomical study revealed that the cause of embryo abortion of wide hybrid *C. baccatum* × *C. annuum* was a result of an early hardening of the endosperm that dramatically limited the embryo's development (Manzur et al. 2015). To overcome the interspecific incompatibility, embryo rescue or a genetic bridge crossing may be employed.

To introgress the anthracnose resistance from *C. baccatum* "PBC81" into a *C. annuum* cultivar, embryos at torpedo or early cotyledonary stage were rescued by being excised from the mother plant and cultured on a medium under sterilized conditions (Yoon et al. 2006). Alternatively, a genetic bridge may be applied by using a species that is close to both *C. annuum* and *C. baccatum* to obtain a hybrid of either target species first, and then the hybrid is crossed again with the other target species. Theoretically, either *C. chinense* or *C. frutescens* could serve as a genetic bridge for the *C. annuum* and *C. baccatum* hybridization. However, Manzur et al. (2015) proved that *C. chinense* was a better genetic bridge than *C. frutescens*. For anthracnose resistance breeding, *C. chinense* was used as a bridge to help introgress *C. baccatum*, either "PBC80" or "PBC81," into *C. annuum*, thus making a trispecies hybrid (Yoon & Park 2005; Temiyakul et al. 2010).

C. baccatum "PBC80" and "PBC81" are resistant to *Colletotrichum scovillei*, the most pathogenic *Colletotrichum* species infecting chili, while *C. chinense* "PBC932" is susceptible. *Co. scovillei* is the predominant pathogen in Korea (Park et al. 2009), Indonesia (Syukur et al. 2013), Taiwan (Liao et al. 2012a; Paul W.J. Taylor unpublished), while *Co. truncatum*

Figure 4.1 Anthracnose-resistant chili genotypes: (1) *Capsicum baccatum* "PBC80," (2) "PBC81," and (3) *C. chinense* "PBC932."

is the main problem in India (Vasanthakumari & Shivanna 2013) and Thailand (Than et al. 2008; Montri et al. 2009). Although "PBC80" and "PBC81" are more resistant than "PBC932," they are not easily hybridized with *C. annuum* compared with "PBC932"; hence, the choice of which resistance sources to use is justified by the predominant pathogen species in the area.

4.2 *Genetics of anthracnose resistance in* Capsicum

Extensive genetic studies have been conducted in several chili populations derived from the main three resistance sources, "PBC932," "PBC80," and "PBC81." These common resistance sources have been introgressed into the local *C. annuum* cultivars to improve the resistance. The crossings yielded chili populations segregating for anthracnose resistance, which were used for genetic analyses of the resistance. In addition, Kim et al. (2008b) identified HR resistance in another *C. baccatum*, "PI594137." Most studies that screened lines for these HR-resistant sources evaluated the resistance on detached and wounded chili fruit. Table 4.1 summarizes the outcomes of the genetic studies carried out in various chili populations derived from the four genotypes of *C. chinense* and *C. baccatum*.

The resistance derived from "PBC932" appeared to be controlled by single recessive genes. The "PBC932"-derived resistance was highly heritable; the resistance expressed in genotypes in the interspecific F2 progenies was as high as in the "PBC932" parent. A resistant *C. annuum* genotype can be achieved by backcrossing to the recurrent parent, which is normally *C. annuum* elite cultivar. For example, the "AR" cultivar, originally developed by the World Vegetable Center, was a BC3F6 progeny derived from "PBC932" (Kim et al. 2007b). On the other hand, a new cultivar with resistance can be achieved via the single seed descent approach. In each generation, the resistant genotypes are selected and self-pollinated to produce further generations. Cycles of selection and self-pollination are repeated until the selected chili lines no longer segregate and thus become pure lines.

C. baccatum "PBC80," "PBC81," and "PI594137" have broader resistance to anthracnose than *C. chinense* "PBC932" (Mongkolporn et al. 2010; Kim et al. 2008b). The genetic studies of anthracnose resistance derived from these *C. baccatum* accessions were reported from intraspecific populations. Wide interspecific hybrid populations generally do not segregate in a Mendelian fashion, which causes difficulty in analyzing inheritance in these populations (Yoon et al. 2009). Inheritance of resistance studied in three different intraspecific *C. baccatum* populations revealed similar

Table 4.1 Genetic analyses of anthracnose resistance derived from various *Capsicum* sources with different resistance assessment

Resistance sources	Chili population[a]	Anthracnose fruit bioassay/ *Colletotrichum* spp.[b]	Resistance mechanism	Genetic mode[c]		References
				Mature green	Ripe	
C. chinense "PBC932"	Intersp. *C. annuum* F2, BC1	Wounded detached fruit/*Ct.**	HR	Single recessive	Single recessive	Pakdeevaraporn et al. (2005), Kim et al. (2008a), Mahasuk et al. (2009a)
	Intersp. *C. annuum* F2, BC1	Wounded detached fruit/*Ca.***	Small lesion	Polygene dominant	Polygene dominant	Sun et al. (2015)
"PBC932"-derived *C. annuum*	Intrasp. F2, BC1	Wounded detached fruit/*Cs.**	Small lesion	Complementary dominant	Duplicate recessive	Lin et al. (2007)
C. baccatum "PBC80"	Intrasp. F2, BC1	Wounded detached fruit/*Cs.**	HR	Single recessive	Single dominant	Mahasuk et al. (2009b, 2013)
C. baccatum "PBC81"	Intersp. *C. annuum* BC1	Wounded detached fruit/*Ca.***	HR	na	na	Yoon et al. (2009)
C. baccatum "PI594137"	Intrasp. F2, BC1	Wounded detached fruit/*Ca.***	HR	Single dominant	na	Kim et al. (2008b)

(*Continued*)

Table 4.1 (Continued) Genetic analyses of anthracnose resistance derived from various *Capsicum* sources with different resistance assessment

Resistance sources	Chili population[a]	Anthracnose fruit bioassay/ *Colletotrichum* spp.[b]	Resistance mechanism	Genetic mode[c]		References
				Mature green	Ripe	
C. annuum cv. "Perennial"	Intrasp. F2, BC1	–/*Ct.*	Small lesion	Additive polygene; na		Ahmed et al. (1991)
C. annuum cv. "Chungryong"	Intrasp. F2, BC1	Wounded detached fruit/*Cd.*** and *Cg.***	Small lesion	Partial dominant	Partial dominant	Park et al. (1990a)
C. annuum cv. "83-168"	Intrasp. F2, BC1	Wounded detached fruit/*Ct.*	Small lesion	Single dominant	na	Lin et al. (2002)

[a] Chili population: Intersp., interspecific; Intrasp., intraspecific.
[b] *Colletotrichum*: *Ca.*, *Co. acutatum*; *Cd.*, *Co. dematium*; *Cg.*, *Co. gloeosporioides*; *Cs.*, *Co. scovillei*; *Csi.*, *Co. siamense*; *Ct.*, *Co. truncatum*.
[c] na, Genetic information due to fruit stages not available.
* The *Colletotrichum* species was revised by multi-gene phylogenetic analysis (Mongkolporn & Taylor 2018).
** Caution: the *Colletotrichum* species identification was not based on multi-gene phylogenetics.

results in that the resistance was controlled by a single dominant gene (Kim et al. 2008b; Mahasuk et al. 2009b, 2013). These populations were made up of a resistant genotype, either "PBC80" or "PI594137," crossed with a susceptible *C. baccatum* parent.

In contrast, a genetic study with another "PBC932" population crossed with a *C. annuum* (Sun et al. 2015) suggested that the resistance was a dominant trait with more than one gene involved. Since the population segregated poorly, the resistance genetics was uncertain. Interestingly, the resistance reaction of the "PBC932" in the study by Sun et al. (2015) was not HR, which could be the reason why the genetics of resistance was different from that seen in previous studies. According to the pathogenicity studies by Mongkolporn et al. (2010), "PBC932" was susceptible to *Co. scovillei*, with the degree of infection varying depending on the pathogen's virulence. A genotype with a small infection may be rated as susceptible in one study while being rated as resistant in another, depending on the difference between the parental genotypes. Parental genotypes are chosen based on a significant difference in their resistance levels so that their progeny population will segregate well for the resistant trait.

In Korea, "PBC81" was introgressed into *C. annuum* via embryo rescue. Its crossability with *C. annuum* was explored with 41 *C. annuum* genotypes. Three of these had partial compatibility, and the hybrid embryos were rescued. Interspecific hybridization incompatibility occurs both pre- and post-fertilization. In this case, there was no problem during pre-fertilization, as the normal pollen germination of "PBC81" was observed on all the stigmas of the pistillate chili parents (Yoon et al. 2006). However, two types of post-fertilization were observed, including embryo abortion and hybrid sterility (for a thorough review on *Capsicum* crossability, see Chapter 2). Successful F1 hybrids were backcrossed to their pistillate parents, and the resistance was recovered in the BC1F1 population.

As well as the HR-resistant genotypes, several previous studies have been conducted in other populations derived from different resistant sources, which were non-HR. These resistances were identified in some cultivars of *C. annuum* and *C. chinense* prior to the discovery of the HR resistance in chili. Partial dominance was identified in a *C. annuum* resistant cultivar (Park et al. 1990a,b). The genetic analyses reported various results, especially the mode of gene action (dominant or recessive) and the number of genes involved. The results of different genetic reports may vary because of differences in resistant sources, inoculation methods, phenotypic classifications, and most importantly, fruit maturity. Table 4.1 summarizes the inheritance of anthracnose resistance derived from various donors, *Colletotrichum* species, and host reactions.

4.3 Differential resistance affected by differential fruit maturity

Resistance to anthracnose in chili is unique in that the assessment is carried out on chili fruit, mostly detached at different fruit maturities. Genetic analyses in different chili populations derived from "PBC932" and "PBC80" revealed that the resistance assessed in mature green and ripe fruit was controlled by different genes (Mahasuk et al. 2009a,b, 2013; Table 4.1). The earliest report on anthracnose assessment at different stages of fruit maturity was in 1990 by Park et al. (1990a,b); studies were conducted in *C. annuum* populations using green and mature red, wounded, detached fruit. However, independent studies 15 years later provided genetic proof of resistance at different fruit stages (Lin et al. 2007; Mahasuk et al. 2009a,b, 2013). This unique differential resistance of different fruit maturity stages appears to be common to all chili populations.

The resistance genes in mature green and ripe fruit in "PBC932" were found to be linked on the basis of gene independence analysis (Mahasuk et al. 2009a). Resistance at the seedling stage was also compared with resistance at the fruit stages. Mahasuk et al. (2009a) identified three recessive genes that conferred resistance at the seedling, mature green, and ripe fruit stage, respectively. The two fruit genes were linked to but independent of the seedling gene. Although anthracnose is not a serious problem for chili seedlings, the linking of the seedling gene to fruit resistance may provide a robust screening method for fruit resistance, thus speeding up the generations.

Mongkolporn et al. (2010) reported that "PBC932" mature green fruit was more resistant than ripe fruit. Among the three *Colletotrichum* species tested, "PBC932" at both fruit stages was resistant (HR) to *Co. siamense* but very susceptible to *Co. scovillei*. Differential host reactions were found in "PBC932" inoculated with *Co. truncatum*. The "PBC932" mature green fruit showed HR resistance to all the *Co. truncatum* isolates, while the ripe fruit showed disease scores of 7–9 when inoculated with *Co. truncatum* Pathotype 1. The details of disease scoring and pathotype identification are described in Chapter 3.

In contrast, "PBC80" expressed resistance in a dominant fashion in the ripe fruit and in a recessive way in mature green fruit (Mahasuk et al. 2009b, 2013). The genes in both fruit stages appeared to be independent. However, in contrast to "PBC932," the "PBC80" ripe fruit was more resistant than the mature green fruit. In general, "PBC80" expressed the broadest resistance to all of the three major *Colletotrichum* species. Except for *Co. scovillei*, differential host reactions were found in the mature green fruit of "PBC80." The mature green fruit showed symptoms with scores 3–5 (2%–15% lesion size as a percentage of fruit size) when inoculated with *Co. scovillei* Pathotype 1, the most aggressive isolate group in Thailand (Mongkolporn et al. 2010).

Molecular studies to support the role of chili fruit maturity in anthracnose resistance were conducted. Host–pathogen interaction was investigated in a *C. annuum* cultivar that showed differential reactions on the mature green and ripe fruit. This particular chili cultivar was susceptible to *Colletotrichum* on the mature green fruit and then became resistant when the fruit ripened. A histochemical examination by transverse sectioning of the inoculated chili fruit found fungal infection 24 h after inoculation through appressoria and infection pegs. More numerous and longer appressoria have been identified as occurring on mature green fruit than on ripe fruit (Kim et al. 1999; Oh et al. 2003); these studies isolated a number of genes that were highly expressed in the ripe chili fruit during fungal infection. One of these was *PepEST* pepper esterase (Kim et al. 2001), a hydrolytic enzyme, which was found to have activity inhibiting appressorium formation.

Differential reactions were also observed in "PBC80" fruit when inoculated with the *Co. scovillei* isolate "MJ8," which belongs to Pathotype 1. Mature green fruit showed anthracnose symptoms, while the ripe fruit showed HR (Figure 3.8; Mongkolporn et al. 2010). This particular host–pathogen interaction was used to study physiological changes of "PBC80" chili fruit during infection (Temiyakul et al. 2012). Differential resistance according to fruit maturity has been an important issue in chili breeding for anthracnose resistance, which potentially hampers breeding success. Accurate selection for the desired trait is always the key to breeding success. In the case of the resistance derived from "PBC80," inconsistent anthracnose symptoms often occurred on the mature green fruit, although the fruit was collected from the same plant. This inconsistency was suspected to arise from the non-uniform fruit maturity of the green stage. During the post-inoculation incubation, non-uniform fruit color within the same plant was observed (Figure 4.2). This suggested that the resistance gene was active in the late mature green stage (Figure 4.3). Figure 4.3 shows fruit color development ("a value") in relation to the anthracnose symptoms at different fruit ages from Weeks 4 to 12. The chili fruit turns red during Weeks 8 and 9, when "a value" becomes positive. Symptom inconsistency occurred if the fruit collected for inoculation was not physiologically uniform. In breeding for anthracnose resistance, fruit maturity has become a fundamental issue for achieving durable resistance selection.

4.4 Breeding for anthracnose resistance via embryo rescue and anther culture

Chili belongs to the *Capsicum* genus, with the popular cultivated species including *C. annuum, C. baccatum, C. chinense,* and *C. frutescens.* However, the most widely cultivated *Capsicum* is *C. annuum,* which

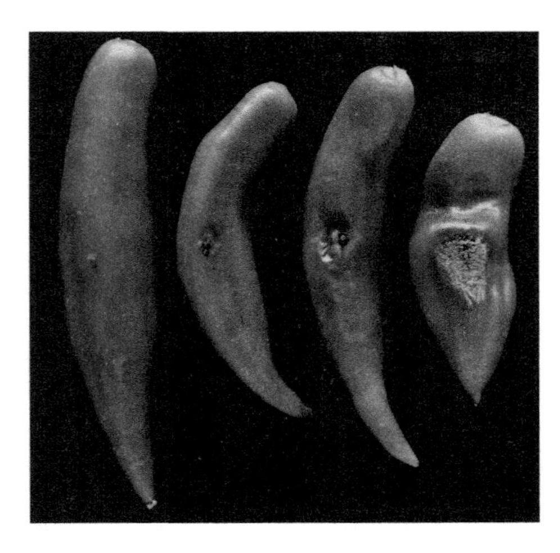

Figure 4.2 Inconsistency of anthracnose symptoms on mature green fruit of *Capsicum baccatum* "PBC80."

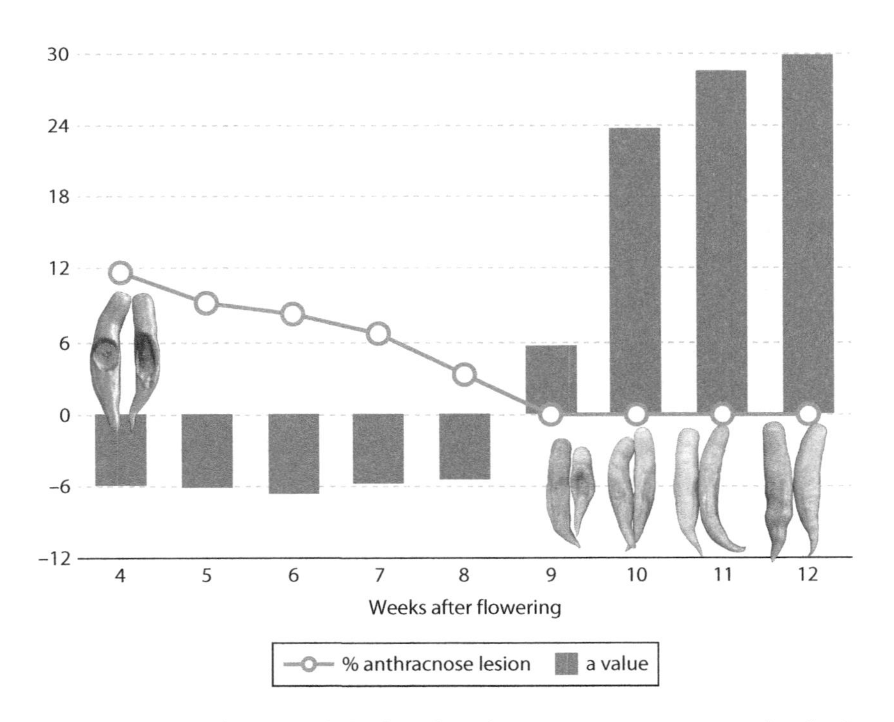

Figure 4.3 (See color insert.) Reduced anthracnose symptoms on developing *Capsicum baccatum* "PBC80" fruit. (From Temiyakul, P. et al., *The Journal of King Mongkut's University of Technology North Bangkok* 22, 494–504, 2012.)

needs improvement in resistance to pathogens and pests. Wide species hybridization is a potential solution to assist with improving traits in *C. annuum* elite cultivars, since anthracnose resistance has been found in *C. baccatum*. Wide hybrids of "PBC80" crossed with some elite *C. annuum* cultivars have been successfully produced with the aid of embryo rescue. However, advanced chili generations leading to pure line development are often hindered by sterile hybrids (Manzur et al. 2015). These hybrids have been backcrossed to the *C. annuum* parent, and the progenies with anthracnose resistance have been selected. The procedure normally takes six to seven generations of backcrossing or selfing to obtain chili pure lines. Alternatively, the backcrossed resistant progenies can be cultured to produce pure lines via double haploid (DH) plant production.

DH plant production is a shortcut to produce a pure line that is completely homozygous in one generation, which makes the DH population ideal for mapping. Another great advantage of DH is that it facilitates the recovery of recessive genes controlling important traits. Chili DH production can be done via anther culture (Irikova et al. 2011). Anther culture simply involves culturing microspores to regenerate haploid embryos; however, the success rate relies heavily on plant regeneration, which seems to be a challenge for chili anther culture and DH production. Among the cultivated *Capsicum* species, *C. annuum* is the easiest species to culture, providing the highest rate of plant regeneration (9.5%; Meekul & Chunwongse 2010).

The first success in plant regeneration from anther culture was reported in *C. annuum* (Wang et al. 1973; George & Narayanaswamy 1973), followed by *C. frutescens* (Novak 1974). Factors affecting the plant regeneration include the genotype, age, and growing conditions of the anther donor plant; microspore developmental stage; and culture media (Irikova et al. 2011). The chili genotype appeared to be the most influential in the success of chili anther culture, which was confirmed by the studies by Raweerotwiboon et al. (2014a,b) and Raweerotwiboon and Chunwongse (2015).

DH chili lines with anthracnose resistance derived from "PBC80" were successfully produced via anther culture (Raweerotwiboon et al. 2014a,b; Raweerotwiboon & Chunwongse 2015). Chili flower buds that contained the majority of the late uninucleate microspores (Figure 4.4) were chosen as the most suitable developmental stage for anther culture and embryo induction (Raweerotwiboon et al. 2014a,b; Barroso et al. 2015). However, chili microsporogenesis was reported to be asynchronous, such that different microspore stages could be found in a single anther (Barroso et al. 2015). Bud stages with the highest number of uninucleate microspores were reported differently in different chili genotypes. For example, the bud stage of *C. annuum* as shown in Figure 4.4 (the length of

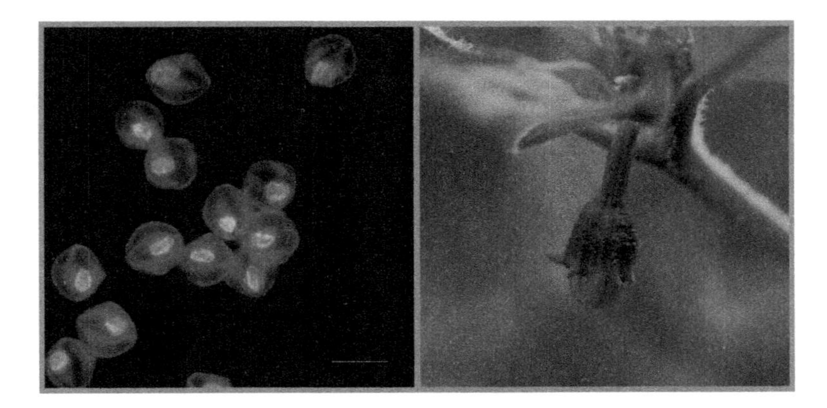

Figure 4.4 (left) Uninucleate microspores stained with 4′-6-diamidino-2-phenylindole (DAPI) fluorochrome; scale bar represents 20 μm. (right) Chili flower bud stage with majority of the late uninucleate microspores. (Microspore photograph courtesy of Dr. Anchalee Raweerotwiboon.)

the petal is longer than the sepal length) was found to be the most suitable (Raweerotwiboon et al. 2014a,b), while the younger bud (petal and sepal lengths are the same) was reported to be suitable in a different chili genotype (Barroso et al. 2015). However, plant regeneration rate is the key to culturing success. All the backcrossed interspecific chili hybrids were directly regenerated to plantlets (direct embryogenesis) at a low rate of 0.38%–0.81%, except for one cross, which had the highest rate of 6.14%. The type of chili genotype within the same *C. annuum* species appeared to play an important role in the success of plant regeneration. In these anther cultures, spontaneous DH occurred at a rate of 1.1–2.5 to haploid plant number. The higher regeneration rate a genotype produced, the higher the DH:H ratio that was obtained.

4.5 Conclusions and remarks

HR resistance to chili anthracnose has been identified in a few cultivars of *C. baccatum* and *C. chinense*, which the most resistant cultivars being "PBC932," "PBC80," and "PBC81." Resistance was characterized by HR and was commonly assessed on detached chili fruit after wound inoculation. Uniquely, the fruit maturity stage played a key role in the differential resistance. Genetic studies identified independent genes conferring the resistance in mature green and ripe fruit. Pathogen–host interaction determined which resistant chili sources were to be used due to the predominant *Colletotrichum* species in the area. Since *C. annuum* is the most popular species but lacks anthracnose resistance, introgression of the resistance from its relatives is encouraged. Interspecific hybridization was achievable from the cross between *C. annuum* and *C. chinense.* Wide hybridization between *C. annuum* and *C. baccatum* was only achieved by embryo rescue. To produce resistant chili pure lines, the wide hybrids have to be backcrossed to the *C. annuum* parent and then successfully developed into DH lines via anther culture.

chapter five

Molecular studies for anthracnose resistance

Although it is a desirable trait, resistance to chili anthracnose is difficult to select due to the complicated host–pathogen interaction. Genetic determination of resistance in *Capsicum chinense* and *C. baccatum* has revealed genes that are differentially expressed at different fruit maturity stages. These genes generate a hypersensitive reaction (HR) in highly resistant genotypes. HR genes have been incorporated into breeding programs to improve commercial chili cultivars, whereby the selection of resistance is through a fruit bioassay. The selection procedure based on the fruit bioassay is laborious, time consuming, and also prone to errors, particularly due to the variation in timing of the development of each fruit maturity stage during the fruit bioassays. Conventionally, disease assessment on each chili plant requires fruit inoculation after wounding the fruit at both mature green and ripe stages. In addition, conventional selection discourages simultaneous selection for several traits. Therefore, a molecular mapping approach has been developed in the anthracnose breeding programs to locate the resistance genes and their linked molecular markers. Successful mapping of the traits in the *Capsicum* genomes will be of great benefit for efficient and accurate selection in combining traits for the future.

5.1 *Molecular marker applications in* Capsicum

Genetic variation in living organisms is essential for evolution, as it provides the opportunity for adaptation to stress. The existence of genetic variation is greatly beneficial to both natural and artificial selection. Some genetic variations are visible through phenotypic expression, although most genetic variation can only be detected through molecular technology. Molecular marker technology plays a significant role in detecting genetic variation and thus is a useful tool for crop improvement. Marker technology can be applied by genetic and association mapping, marker-assisted selection (MAS), and genetic diversity.

5.1.1 *Mapping and quantitative trait loci (QTL) analysis*

5.1.1.1 *Genetic mapping*

An important goal for crop improvement is to locate the gene(s) of interest. Knowing the location of the genes enables the identification, cloning, and manipulation of the genes. Molecular markers linked to the genes can be used to select the trait, thus enabling more efficient and accurate selection. Mapping is an approach for locating genes on chromosomes that analyzes genetic linkage among molecular markers and genes across the genome. Markers and genes that are linked do not independently segregate in a Mendelian fashion but form a linkage group in linear orders representing a chromosome or part of a chromosome, which is based on their likelihood of co-inheritance (Hartl & Jones 2005). Linkage, or genetic distance, is calculated in map units called *centimorgans* (cM) based on recombination frequencies between the markers and the genes residing on the same chromosome.

A genetic map aiming to locate the gene of interest is theoretically constructed from a segregating population derived from a single cross, such as F2, BC1, or BC2. Ideal parents should be highly homozygous but genetically diverse at the targeted trait, such as disease resistance. The mapping population thus segregates for the trait of interest, which will enable the association of the phenotyped trait and the mapped molecular markers to be analyzed. Consequently, the positions of the genes controlling the trait in the genome can be located.

Proper population size depends on the complexity of the trait studied. Ferreira et al. (2006) recommended 200 individuals as a sufficient number for mapping a simple trait based on a comparison study using four types of population, i.e., F2, BC1, recombinant inbred lines (RIL), and double haploid (DH). Insufficient individuals in the mapping populations resulted in inaccurate maps; the larger the population size, the more precise the map. The degree of inaccuracy also depended on the population types. Heterozygous populations such as F2 and BC require a larger population than RIL and DH; however, maps constructed from heterozygous populations can be improved with co-dominant markers. The ideal population type for mapping should be homozygous at all loci, which can be derived from RIL and DH. The biggest advantage of the RIL and DH over the F2 and BC populations is that they are genotypically fixed and completely reproducible. However, some limitations apply; the RIL generally takes at least six or seven generations to obtain, while the DH is not available in all crops. In *Capsicum,* the success of DH production is highly specific to the chili genotype (Raweerotwiboon et al. 2014b).

Figure 5.1 shows a production diagram of the four basic mapping population types derived from the same cross. P1 and P2 are parents that are different at the trait to be mapped, which are crossed to generate an F1. The easiest and fastest procedures to obtain segregating populations

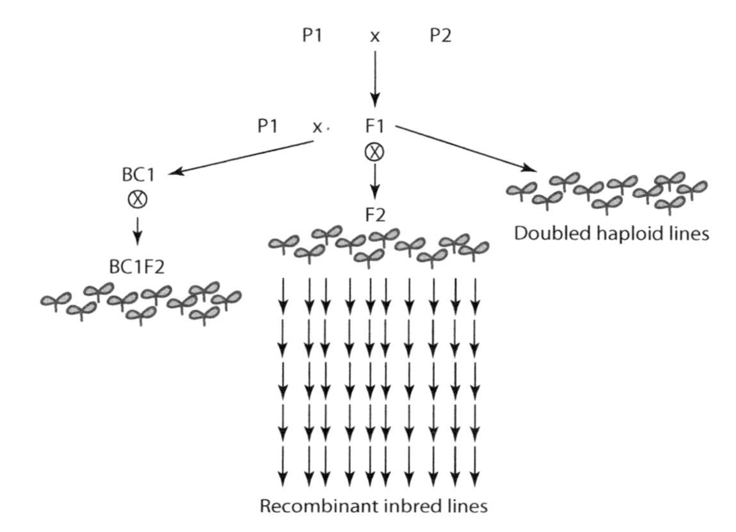

Figure 5.1 Basic mapping populations produced from a single cross of two parents, P1 and P2, include F2, BC1F2, doubled haploid, and recombinant inbred lines.

for mapping are F2 and BC1, which are derived from selfing the F1 and backcrossing F1 to one of the parents, respectively. F2 and BC1 are highly heterozygous, which could be disadvantageous for mapping compared with RIL and DH. RILs contain a series of homozygous lines originated by inbreeding every single F2 individual.

5.1.1.2 QTL analysis

In the case where a trait is determined by multiple genes or is quantitative (QTL), typically, the phenotypic distribution is continuous and does not fit a single gene model (Figure 5.2). Usually, the phenotypic expression of the QTL is substantially affected by the environment; therefore, the heritability of the trait is less than 100% (Prince et al. 2013). QTL traits in *Capsicum* are yield, fruit size, color and shape, and capsaicinoid content. QTL traits are too complicated for conventional selection, although with the aid of molecular markers, QTL selection has become more achievable (Collard et al. 2005; Prince et al. 2013). Based on the constructed genetic linkage framework, QTL analysis is able to detect the locations of the QTLs that influence the phenotypic variation. Three common methods to detect QTLs include single point analysis, simple interval mapping, and composite interval mapping (Tanksley et al. 1996; Collard et al. 2005; Prince et al. 2013). Single point analysis is the simplest, using linear regression to identify the correlation between a marker and the phenotypic variation. This approach does not require a complete linkage map to perform the

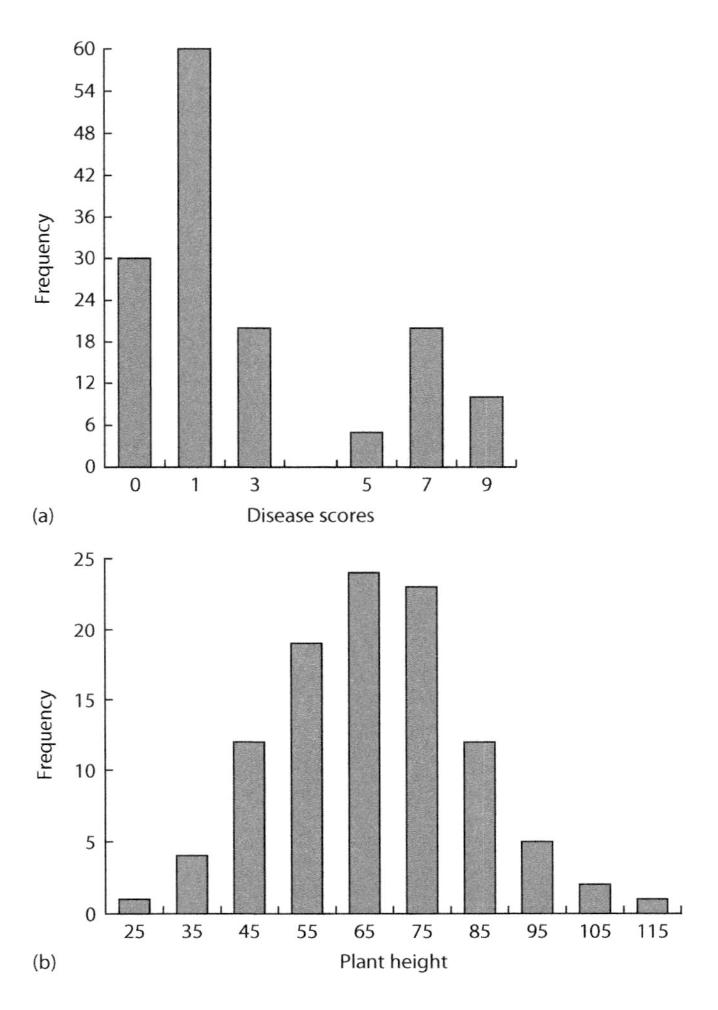

Figure 5.2 Phenotypic distribution based on a single gene model (a) and a QTL (b).

basic statistical regression analysis. However, this method is limited to a close distance between the marker and the QTL, normally at intervals less than 15 cM (Tanksley 1993).

Interval QTL analysis, also called *simple interval mapping* (SIM), analyses intervals between adjacent pairs of linked markers instead of single markers. SIM is based on maximum likelihood, assuming that a single QTL resides between flanked markers, which compensates for recombination between the markers and the QTL (Paterson et al. 1988). Composite interval mapping (CIM) was developed later by two independent research groups (Jansen & Stam 1994; Zeng 1994). CIM combines SIM and linear regression and includes unlinked markers in the analysis in addition to the flanking markers. Consequently, CIM provides a higher-resolution map with more precise and effective detection of true QTLs with small effects (Collard et al. 2005; Prince et al. 2013).

5.1.1.3 Physical mapping

A physical map is based on actual DNA nucleotides or base pairs and is not directly related to the genetic map, which is based on calculated recombination frequency. Similar genetic distances from different genomes or on different chromosomal segments can be physically different. This is due to the fact that recombination occurs more frequently in some regions than in others. A physical map is derived from a high-resolution map. The high-resolution map is a genetic map, partially constructed or covering the whole genome, which contains highly dense markers, 1 cM apart on average. The known sequence and location of the flanking markers help assemble the contiguous DNA sequences (contigs) into chromosomes by physically mapping the marker sequences on the DNA contigs and comparing this with their known relative locations on the genetic map. Currently, advanced molecular marker technology, such as next generation genome sequencing (NGS), makes it feasible to achieve high-resolution maps. When the markers are physically mapped, the intervening region can be analyzed, which enables identification of the candidate genes. A physical map lying within 1 cM can include several thousands to millions of base pairs, especially for a large and complex genome such as *Capsicum*. Table 5.1 reveals that 1 cM of the *Capsicum* genome ranges from 900 Kb to 3 Mb (Han et al. 2016a,b). Therefore, the region can house hundreds of candidate genes.

Table 5.1 *Capsicum* physical and genetic maps

Chromosome	Genetic distance (cM)	Physical distance (Mb)	Mb/cM
P1	208.5	272.6	1.3
P2	107.5	171.1	1.6
P3	118.5	257.9	2.2
P4	116.5	222.5	1.9
P5	100.6	233.4	2.3
P6	102.6	236.9	2.3
P7	92.5	231.9	2.5
P8	153.7	144.8	0.9
P9	86.2	252.7	2.9
P10	103.9	233.6	2.2
P11	86.8	259.7	3.0
P12	94.9	235.7	2.5
Total	1372.2	2752.8	2.0

Source: Modified from Han, K. et al., *DNA Res.*, 23, 81–91, 2016.

5.1.1.4 Shortcut to identify markers

Mapping and QTL analysis provides the locations of markers closely linked to the targeted gene(s), which can be used to routinely screen for the trait. However, the linked markers do not necessarily need to be identified via mapping procedures. A classical approach was developed 26 years ago to rapidly identify molecular markers linked to a gene conferring resistance to powdery mildew in lettuce (Michelmore et al. 1991). The technique was called *bulked segregant analysis* (BSA), which involves screening for molecular differences between two DNA bulks. The DNA bulks originate from a population segregating for the trait of interest, which normally is F2 or BC. The key feature of the DNA bulk is that members of each DNA bulk are phenotypically similar, i.e., resistant or susceptible. The two DNA bulks are consequently genetically different at the specific gene region, while other regions are arbitrarily heterozygous (Figure 5.3). The DNA pooling trick enables rapid identification of the linked markers by screening for DNA polymorphisms between the two bulks derived from early-generation populations such as F2 and BC.

Once the linked markers are identified, the markers can be routinely used to select traits of interest without the need to evaluate the phenotype. Therefore, MAS is advantageous in breeding programs by increasing the accuracy of selecting difficult traits (such as those that require a specific time or location to perform the selection or have low heritability); eliminating unreliable phenotypic evaluation; accelerating selection time, such as at the seedling stage; eliminating genetic drag or linked undesirable trait; and enabling gene pyramiding.

5.1.2 Genetic diversity

Broadly, the term *genetic diversity* refers to variations in DNA, genes, and traits within a population. Traditionally, crop diversity has been analyzed using morphological traits of interest, which are limited. Molecular markers such as random amplified polymorphic DNA (RAPD), amplified fragment length polymorphism (AFLP), simple sequence repeat (SSR), and single nucleotide polymorphism (SNP) are therefore used to measure genetic variation. The study of genetic diversity is valuable for clarifying evolutionary relationships and taxonomies as well as providing an understanding of genome changes within and between species. Understanding the genetic diversity of crops will scope out a crop breeding strategy (Hayward et al. 2015), such as heterosis by intraspecific hybridization or wide hybridization. Crop breeding is generally aimed at increasing yield and enhancing quality, adaptability, and resistance to biotic and abiotic stresses. Global climate changes have forced crop breeding to produce more novel traits for sustainable agriculture. To meet the challenges, plant

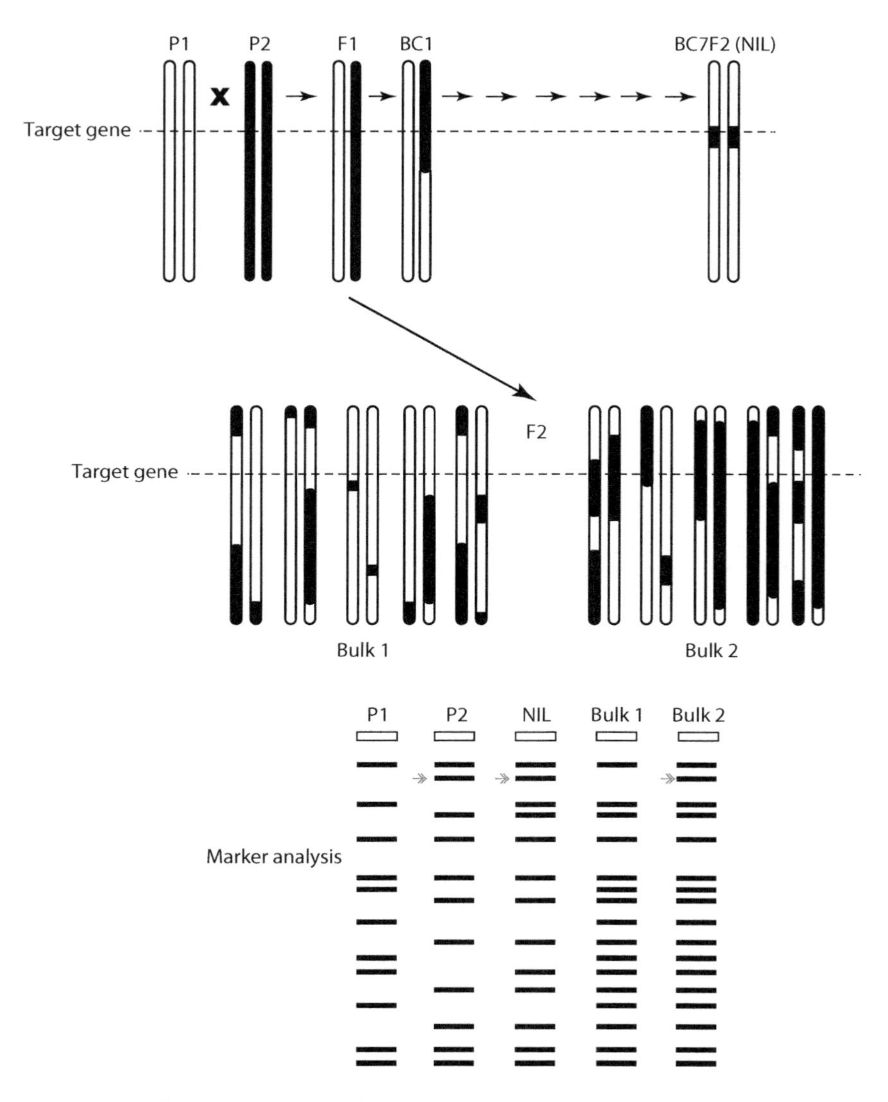

Figure 5.3 Bulked segregant analysis (BSA): recurrent (P1) and donor (P2) parents are crossed to generate F1, aiming to introgress the target gene from P2 into P1 through backcrossing. BC7 is a near isogenic line (NIL) of P1 and contains the target gene from P2. BSA helps to identify markers linked to the target gene as early as F2 generation. The identified marker in the F2 is instantly applied to select the BC progenies containing the target gene in the NIL production. Arrows indicate the marker that links to the target gene. (Modified from Giovannoni, J.J. et al., *Nucleic Acids Res.*, 19, 6553–6558, 1991.)

breeders need to increase their efficiency and exploit crop genetic diversity more thoroughly (Fu 2015).

SSRs and SNPs are widely used in crop genetic diversity analysis due to their abundance and hyper-variability. In particular, SNPs are often free of selective pressures, which enables a complete estimate of diversity levels based on random genetic drift (Hayward et al. 2015). Therefore, SNPs are very useful for tracking chromosome segments to identify recombination events and break up the regions. Molecular markers are also helpful in germplasm management, not only for the estimation of genetic diversity but also for identification, the elimination of duplicates, the protection of elite genotypes (Kesawat and Das 2009), and establishing a core collection (Mongkolporn et al. 2015).

5.1.3 *Association mapping*

Association mapping refers to the statistical analysis of associations between molecular markers and traits of interest in a germplasm collection (Hayward et al. 2015). Molecular markers linked to traits under selection contribute to phenotypic variation based on linkage disequilibrium (LD). LD refers to the co-inheritance of specific markers in related individuals at higher frequencies than expected based on recombination distances. Association analysis can be based on either candidate genes or a whole genome. The candidate gene approach aims to determine correlations between traits and molecular markers within candidate genes presumed to be involved in the traits. The whole-genome approach aims to identify potential LD-associated loci across the genome with densely mapped markers. Association mapping has become popular for mining new alleles using trait–marker relationships within species of natural germplasm collections, where the production of large biparental mapping populations is not feasible.

Capsaicinoid content and fruit weight are traits resulting from *C. annuum* domestication. Genome-wide association analysis (GWAS) was able to identify SNPs associated with both traits (Nimmakayala et al. 2016a). The study generated 66,960 SNPs using genotyping by sequencing (GBS) from a germplasm collection of 96 *C. annuum* accessions representing worldwide geographical areas. Several identified SNP loci were located in the known genes involved in capsaicin production and fruit weight. In the *C. baccatum* germplasm collection, GWAS identified 36 SNPs associated with the peduncle length trait (Nimmakayala et al. 2016b). Peduncle length is an important trait that differentiates wild and domesticated *C. baccatum*.

Figure 5.4 represents an example of a GWAS study for capsaicin content in a collection of 250 Thai chili landraces. The study developed over 22,000 genome-wide SNPs from *Capsicum* using the DArTseq technique.

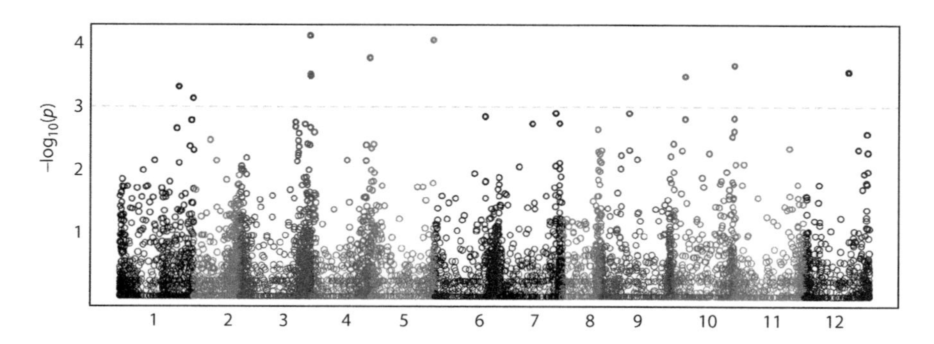

Figure 5.4 Manhattan plot of the genome-wide association study on capsaicin content in Thai chili landraces. Each dot represents a SNP locus. Chromosome coordinates are on the x-axis, and $-\log_{10}$ of the association *P* value for each SNP is on the y-axis. High $-\log_{10}$ (*P*), above 3, suggests strong association with the trait. (Plot courtesy of Wassana Kethom.)

Approximately 9,600 SNPs were analyzed for capsaicin content association using an efficient mixed model. Nine SNPs located on chromosomes 1, 3, 4, 5, 10, and 12 were associated with the capsaicin content.

5.1.4 *Comparative genomics*

The comparative study of genomes is an important approach to draw inferences about the function and evolution of organisms. The complete sequences of several genomes have shed some light on genome structure and gene repertoire, although it is still not enough to understand all the important phenotypes. Identifying all the functional sequences and using them to improve crops is the next task for researchers. The alignment of DNA sequences of multiple organisms is the core process in comparative genomics (Hardison 2003) by finding the parts of the genome of crop species that show similarities in their DNA profiles. The *C. annuum* genome has been sequenced (Qin et al. 2014); therefore, other *Capsicum* species can now be compared. Several taxonomic groups of crop plants that share the same basic chromosome number appear to have similar chromosome maps. After the repetitive DNA is removed, the maps of single copy sequences are syntenic (Jones et al. 2009), indicating that the genes in different crop species are basically the same.

Comparative genomics provides invaluable information regarding evolutionary consequences, including the rate of evolution, gene loss or retention, and duplications and chromosomal rearrangement (Caicedo & Purugganan 2005). The information contributes to taxonomic, morphological, and physiological variations. The sequence availability of genomes coupled with complementary bioinformatics tools helps in redrawing plant phylogeny, identifying ancestral gene sets of crop plants, clarifying the subsequent evolutionary history of these genes, and providing new insight into the causes and consequences of fluctuating genome size. A better understanding of relationships among different plant genomes and their constituent genes will expedite goals ranging from the isolation of genes and determination of their functions to the identification of molecular markers useful for MAS selection in breeding programs (Rahman & Paterson 2009).

5.2 *Molecular marker technology*

A molecular marker is a DNA fragment with or without known sequences that differentiates two or more individuals within a population. The DNA differences are caused by mutations, i.e., DNA deletion, insertion, and substitution. Individuals differentiated by a marker are not necessarily phenotypically different and can be simply visualized by electrophoresis. Molecular marker technology has evolved dramatically in the past decade and is currently employing NGS. Molecular marker technology can be classified into three developmental phases.

5.2.1 *First-generation markers*

The markers in this generation are based on DNA hybridization techniques. The classic and best-known marker is restriction fragment length polymorphism (RFLP), developed by Botstein et al. (1980) to construct a human linkage map. The basics of RFLP development involve (1) DNA digestion: genomic DNA is completely digested by a restriction enzyme, resulting in plenty of DNA fragments of various sizes; (2) DNA fragment separation and blotting: the digested DNA fragments are separated by electrophoresis and transferred onto a nylon membrane for DNA hybridization; (3) DNA hybridization: this is performed on the membrane with a labeled DNA probe; and (4) marker detection: this proceeds after the hybridization is complete. An RFLP marker shows polymorphism based on different lengths of DNA fragments on the same locus; thus, the marker has a co-dominant nature. The fragment length polymorphism arises from mutations that alter the restriction sites. The technique is not currently popular because of its lengthy and complicated procedure.

5.2.2 *Second-generation markers*

Markers developed in the second generation involve polymerase chain reaction (PCR) technology, which amplifies DNA *in vitro* and was developed by Saikai et al. (1985). PCR takes advantage of a thermostable DNA polymerase, isolated from *Thermus aquaticus*, to synthesize new strands of DNA in a single reaction tube. The PCR imitates natural DNA replication by using primers or short DNA sequences to target a DNA segment to be amplified. Therefore, the primer plays a key role in the PCR amplification. Various PCR techniques have been developed based on different types of primers.

RAPD is developed in the early stages by using a short single primer that contains 10 random nucleotides (Williams et al. 1990). By chance, the RAPD primer finds complementary sequences in the targeted DNA and amplifies the primed DNA segments to millions of copies. Generally, the RAPD primer can only amplify DNA fragments for size ranges between 200 and 2000 bp, with priming occurring once in every million bp (Jones et al. 2009). Polymorphism in RAPD is a result of DNA sequence variation in the priming sites, which consequently affects whether the amplification is successful or not. Therefore, RAPD is a dominant marker based on the presence/absence of the amplified DNA fragment and polymorphisms in sequence lengths between primers.

AFLP is a marker technique that uses PCR to amplify restriction DNA fragments (Vos et al. 1995). Double digestion of DNA is applied by using two restriction enzymes: rare and frequent cutters. Both are restriction enzymes with different sizes of recognition sequences. Originally, *Eco*RI, with a six-base recognition site, was used as the rare cutter, and *Mse*I, with a four-base recognition site, was used as a frequent cutter. The combination of rare and frequent cutter enzymes is beneficial for adjusting the number of fragments to be amplified. By using the frequent cutter alone, small DNA fragments are generated. The rare cutter helps reduce the number of DNA fragments. DNA restriction fragments are then ligated with adapters that are compatible with each end and serve as primer binding sites for subsequent amplification. The ligated DNA fragments are first preamplified with an AFLP primer pair with a single selective nucleotide; subsequently, the preamplified DNA serves as a DNA template for the second amplification, which uses longer selective nucleotides as primers. A brief AFLP procedure is illustrated in Figure 5.5.

An SSR or *microsatellite* is repetitive DNA that contains a tandemly repeating unit of simple sequence (1–10 bp) (Tautz & Renz 1984): for example (AG)n, (CT)n, (TGT)n, (TTC)n, (GACT)n, where n is the number of repeats for each SSR locus. SSRs are abundant and highly variable in plant genomes (Powell et al. 1996) due to mutation affecting the number of repeating units. In the *Capsicum* genomes, an SSR is found in every 3.8 kb of expressed sequence tags (Yi et al. 2006). PCR can detect differences among individuals with a primer pair based on the flanking regions of the SSR sequence to be amplified (Figure 5.6). The variations of amplified DNA fragments among individuals arise from different numbers of repeats, resulting in markers with length polymorphism. SSRs have an advantage in that they are co-dominant and highly reproducible as well as species specific.

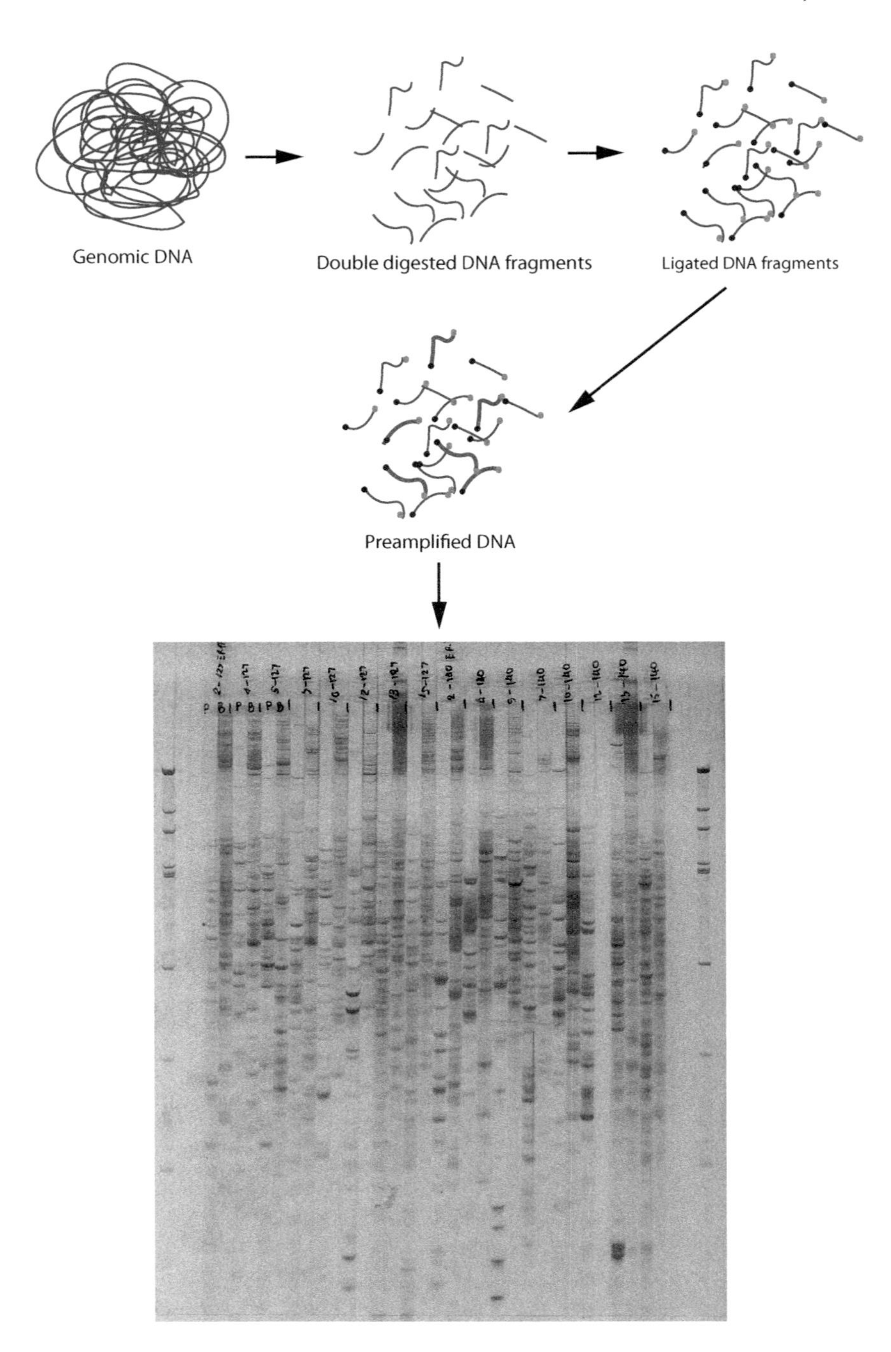

Figure 5.5 Amplified fragment length polymorphism (AFLP) technique (Vos et al. 1995). Adapters with different shades are compatible with different restriction sites.

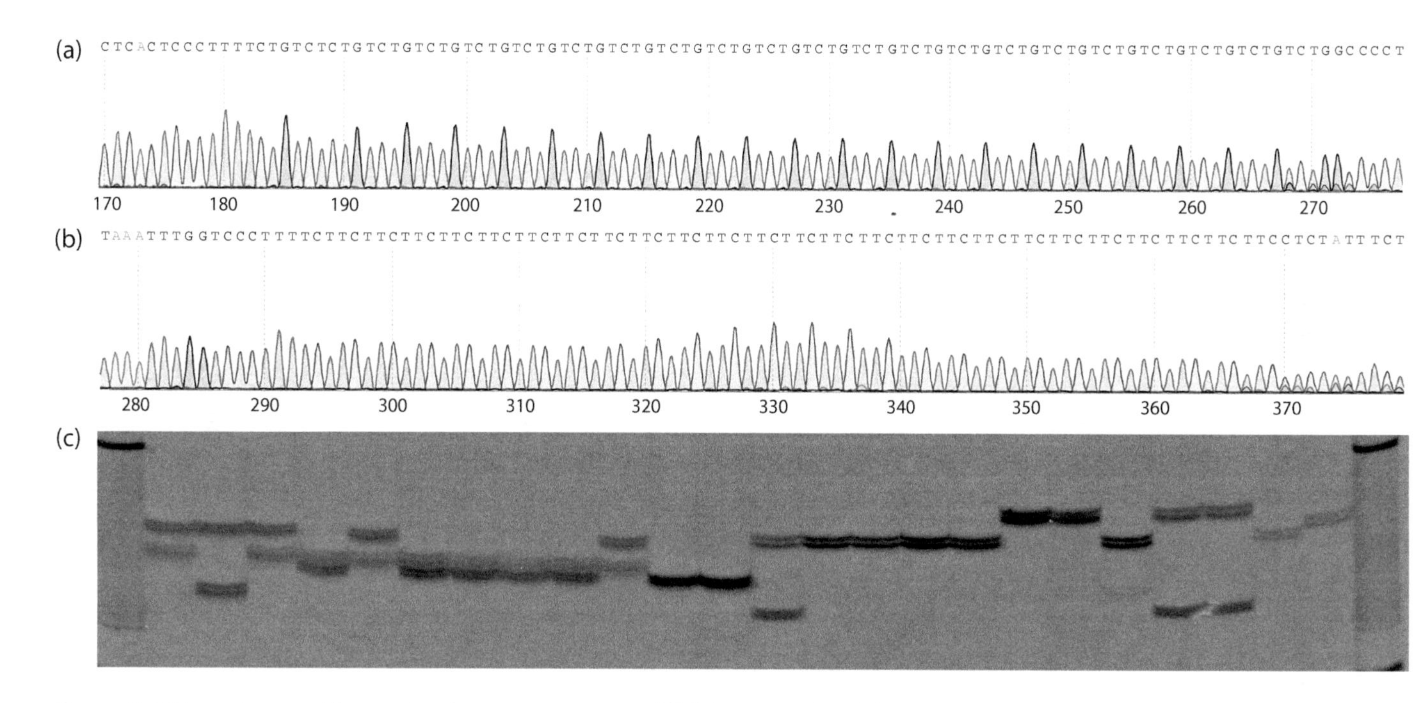

Figure 5.6 Two microsatellite or simple sequence repeat (SSR) loci developed from *Capsicum*: "GTCT" located at bases 191–270 (a) and "CTT" at bases 294–368 (b). DNA sequences flanking the SSR are used to design primers to amplify DNA from chili germplasm. Five allelic DNA fragments of an SSR are visualized in a polyacrylamide gel (c). (SSR chromatograms courtesy of Ruvini Lelwala; polyacrylamide gel photograph courtesy of Dr. Ratri Boonruangrod.)

SSRs are powerful markers for analyzing the genetic diversity of germplasm. Mongkolporn et al. (2015) proved that as few as 10 SSRs were sufficient to study the genetic diversity of 230 chili accessions. However, the SSRs have to be highly informative, based on high polymorphic information content (PIC) values. The SSRs used in the study were selected across the *Capsicum* genome based on their known locations on the *Capsicum* map, with PIC values greater than 0.5.

Capsicum SSRs were developed by two approaches from *Capsicum* genomic libraries and the GenBank database (Lee et al. 2004). By using $(AT)_{15}$, $(GA)_{15}$, $(GT)_{15}$, $(ATT)_{10}$, and $(TTG)_{10}$ as probes, over a hundred clones containing SSRs were isolated from the genome libraries, which led to 40 reliable SSRs being developed and mapped. On the other hand, 36 SSRs were developed from the database. The study showed that the PIC values obtained from the genomic SSRs were double those from the database. These anchored SSRs are useful for other *Capsicum* research, such as gene mapping and diversity analysis.

An inter-simple sequence repeat (ISSR) is a marker originating from variations of DNA lying between two SSR loci (Zietkiewicz et al. 1994). For example, $(CA)_8$ can be used as a primer to amplify variable DNA regions with proper lengths for PCR between two SSR loci complementary to the $(CA)_8$.

A sequence tagged site (STS) is a marker identified by a known and unique DNA sequence 200–500 bp in length (Olson et al. 1989). A primer pair is designed from the borders of the targeted DNA. STSs represent physical landmarks in the genome and are used to tag the larger region of the genome.

A sequence characterized amplified region (SCAR) is a marker developed from multi-locus markers such as RAPD and AFLP markers. A long, 22- to 24-mer primer is designed based on each border sequence of the specific RAPD or AFLP marker to yield a single PCR product corresponding to the original marker sequence. SCAR was originally developed to overcome the low reproducibility of RAPD (Paran & Michelmore 1993).

Cleavage amplification polymorphism (CAP) and single-stranded conformation polymorphism (SSCP) are markers further developed to differentiate PCR products of the same size that are amplified from the same PCR primer pair. These similarly sized DNA fragments often contain a few nucleotide differences that can be differentiated by digestion with a restriction enzyme (CAP; Konieczny & Ausubel 1993) or gel electrophoresis (SSCP; Orita et al. 1989). As a result of DNA digestion, the DNA fragments may or may not be cut at the same sites; thus, DNA fragments of different sizes are expected. On the other hand, two similarly sized DNA fragments that contain different sequences can be differentiated by their secondary structures or the conformation of their single DNA strands.

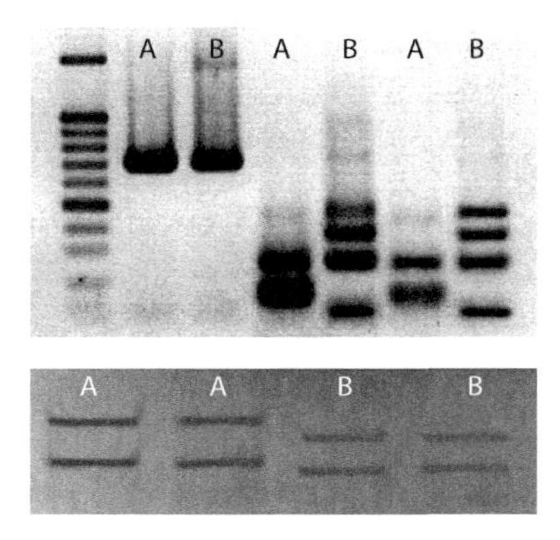

Figure 5.7 Cleavage amplification polymorphism (CAP, top) and single-stranded conformation polymorphism (SSCP, below). Genotypes A and B originally have similar DNA fragment sizes. After DNA digestion with a restriction enzyme, two CAPs are generated. CAP-A contains two DNA bands, and CAP-B contains four DNA bands. Without digestion, Genotypes A and B can be differentiated by SSCP.

The key feature in detecting SSCP is the separation of single-stranded DNA fragments in polyacrylamide gel electrophoresis, whereby different conformations of DNA of the same molecular weight migrate at different speeds. Figure 5.7 shows CAP and SSCP derived from two different genotypes, A and B, with originally similar PCR product size.

5.2.3 Third-generation markers

Markers in this generation are based on DNA sequencing, which mainly focuses on SNP. An SNP is a single base difference between two DNA individuals caused by nucleotide substitutions, including base transitions (C/T or G/A) or transversions (C/G, A/T, C/A, or T/G) (Jiang 2013). In practice, single base variations derived from cDNA (mRNA) are SNPs caused by single base insertions and deletions called *InDel*. SNPs are the most common DNA variations in eukaryotic genomes (single base differences), abundant and uniformly distributed across a genome at an average frequency of one SNP in every 100–300 bp (Metzker 2005; Edwards et al. 2007; Seeb et al. 2011). This high frequency makes SNP markers the most abundant polymorphic markers (Appleby et al. 2009), providing many opportunities to find markers close to genes of interest. Consequently, SNPs have become the markers of choice to map traits of interest (Jones et al. 2009). However, SNP discovery relies on genome sequence information, which can be obtained from public sequence databases (usually model genomes) or newly generated by some sequencing technology (for non-model genomes).

5.2.3.1 SNP discovery

Before NGS was developed, SNP discovery relied on the resequencing of unigene-derived amplicons (Ganal et al. 2009) using low-throughput Sanger sequencing or chain termination (Sanger et al. 1977) followed by *in silico* sequence alignment with reference sequence databases. Amplicon resequencing is costly, laborious, and most importantly, unable to discover SNPs in low-copy coding and non-coding regions (Ganal et al. 2009). Genome-wide SNP discovery in *Capsicum* was achieved by comparing whole-genome resequencing data from two chili cultivars with the reference *Capsicum* genome (*C. annuum* cv. "CM334"). Data mining revealed 6.8 and 4.2 million SNPs for both cultivars (Ahn et al. 2016).

The emergence of NGS has brought high expectations of fast SNP discovery for non-model crops (which lack genome sequence reference databases). NGS is a very high-throughput sequencing technology developed by several companies into different platforms, such as 454 Life Sciences by Roche Applied Science, HiSeq by Illumina, and Ion Torrent by Life Technologies Corporation (Metzker 2010). These technologies can operate up to several billions of bases per run (Behjati & Tarpey 2013), thus allowing SNP discovery of the whole genome at significantly low cost. Sequencing technologies are still evolving; therefore, NGS is now regarded as second generation. New third-generation technology can directly sequence single DNA molecules and eliminate the DNA amplification step, thus resulting in cost reduction (Pillai et al. 2017). Newer technology is also being developed, regarded as fourth generation. Spatial transcriptomics can read nucleic acid composition directly in cells or tissues.

For large-genome crops such as *Capsicum*, before its whole-genome sequence (WGS) completion, large-scale expressed sequence tags (ESTs) or transcriptomes were sequenced with NGS to substitute for the WGS to identify SNPs (Ashrafi et al. 2012; Nicolai et al. 2012; Kang et al. 2014). ESTs consist of a few hundred base pairs of cDNA developed from expressed sequences (mRNA) derived from specific tissues. Sequencing the transcriptome from *Capsicum* enabled tens of thousands of SNPs to be identified while reference sequences were not yet available (Berthouly-Salazar et al. 2016). Figure 5.8 shows an SNP discovered by sequencing.

5.2.3.2 SNP detection and genotyping

SNP genotyping is aimed at distinguishing allelic variations of SNPs among individuals in a population. Several techniques have been developed to perform large-scale genotyping, which have multiplex capacity and high accuracy. SNP detection relies on four basic mechanisms (Sobrino et al. 2005): allele-specific oligonucleotide (ASO) hybridization, primer extension, oligonucleotide ligation, and invasion cleavage.

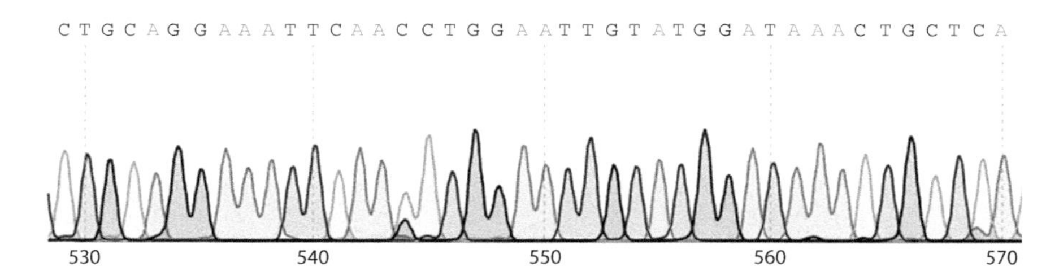

Figure 5.8 (See color insert.) An SNP, C/G, is discovered at base 544. (DNA sequence chromatogram courtesy of Dr. Ratri Boonruangrod.)

- *ASO hybridization* is based on discriminating two DNA samples differing at one nucleotide position by hybridization. Two allele-specific probes are designed, around 20 bp long. Perfectly matched probes hybridize one of the two alleles of a SNP locus. The resulting hybrid DNA remains stable when the temperature is raised, while the non-hybridized strand is unstable (Figure 5.9). The system can be applied on a high-throughput microarray to perform large-scale SNP genotyping simultaneously.
- *Fluorescence resonance energy transfer (FRET)* is a quantum phenomenon that occurs when two fluorescent dye molecules are in close proximity to each other (Didenko 2001). Excitation is transferred from a donor molecule to an acceptor fluorophore. The donor fluorescence is quenched, and the acceptor becomes excited and fluoresces. FRET combined with ASO, which requires two oligonucleotide fluorophores in a quantification real-time PCR, can form a FRET probe. There are several variations in FRET probe designs from different companies.

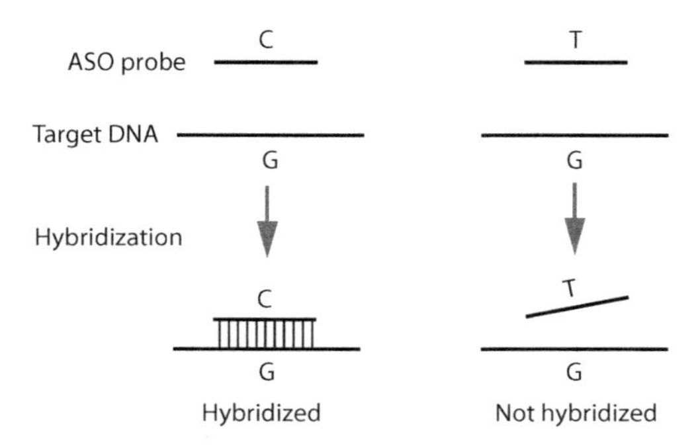

Figure 5.9 Allele-specific oligonucleotide (ASO) hybridization. Two ASO probes are hybridized with the target DNA, which contains an SNP. Only perfectly matched target–probe hybrids are stable. (Redrawn from Sobrino, B. et al., *Forensic Sci. Int.*, 154, 181–194, 2005.)

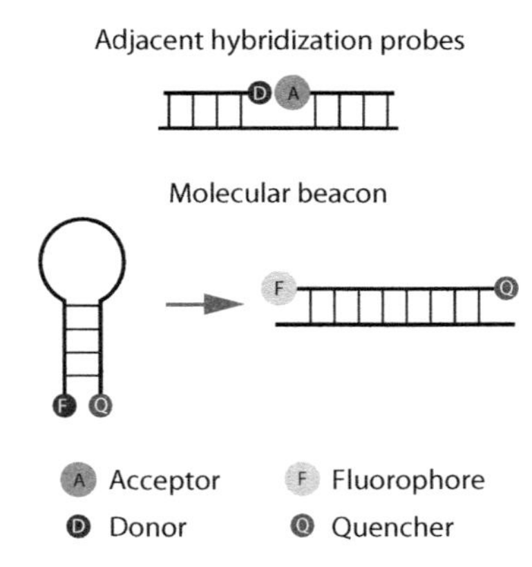

Figure 5.10 FRET probes for hybridization detection. The fluorescence of the acceptor fluorophore is enhanced when two adjacent probes hybridize to the target sequence. Successful probe hybridization changes the configuration of the molecular beacon. The separation of the fluorophore from the quencher restores fluorescence. (Redrawn from Didenko, V.V., *BioTechniques*, 31, 1106–1121, 2001.)

- A *molecular beacon*, a FRET derivative, is a single-stranded oligonucleotide probe with complementary regions at each end, allowing the probe to form a hairpin-like structure (Didenko 2001) (Figure 5.10). A fluorophore is attached at one end of the beacon and a fluorescence quencher at the other. When the beacon probe is not hybridized to the target DNA, the donor and acceptor molecules are brought into close proximity, resulting in the FRET-based quenching of the donor. When the target DNA is hybridized, the beacon probe forms longer and stronger hybrid molecules. The FRET is disrupted, and the donor fluoresces. The fluorescent acceptor fluorophore in molecular beacons is replaced with a non-fluorescent quencher. This detection mechanism is simplified, because the non-hybridized molecule does not fluoresce. Advanced forms of molecular beacons with multiple colors have been designed to be used in multiplex PCRs.
- In *array hybridization*, short oligonucleotides are attached to a solid substrate in a microarray format and are hybridized with fluorescence-labeled PCR products containing the SNP sequence. Large numbers of SNPs are detected in parallel (Sobrino et al. 2005).
- *Primer extension* is based on the capability of DNA polymerase to incorporate specific deoxyribonucleotides complementary to the sequence of template DNA. In a minisequencing reaction, a primer that anneals to the target DNA immediately adjacent to the SNP is extended by a DNA polymerase with a single nucleotide complementary to the polymorphic site (Sobrino et al. 2005) (Figure 5.11). The procedure can be performed in a microarray format.

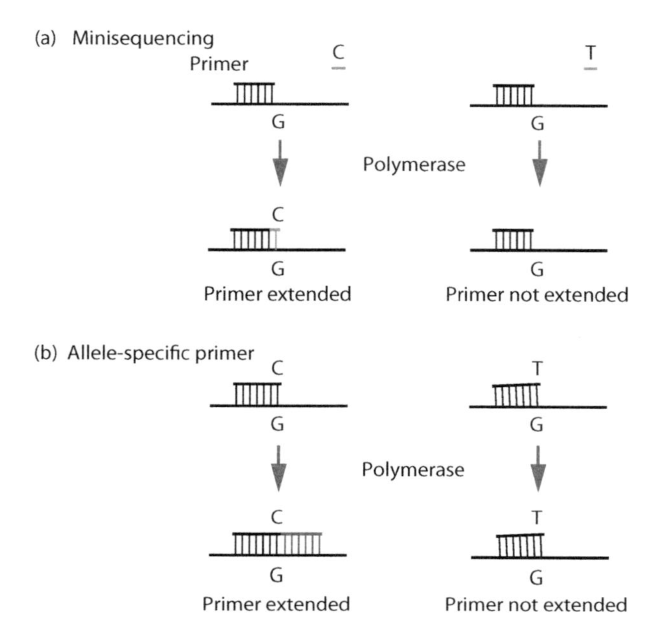

Figure 5.11 Primer extension. Minisequencing (a): a primer anneals to the target DNA immediately upstream of the SNP and is extended with a single nucleotide complementary to the polymorphic base. Allele-specific extension (b): the 3′ end of the primers is complementary to each allele of the SNP, and the primer extends when a perfect match is found. (Redrawn from Sobrino, B. et al., *Forensic Sci. Int.*, 154, 181–194, 2005.)

- *ASO ligation* employs DNA ligase to covalently join two oligonucleotides when they hybridize next to one another on a template DNA (Sobrino et al. 2005). The assay requires three probes, one common and two allele specific. The common probe anneals to the target DNA immediately downstream of the SNP. One allelic probe is complementary to one allele at the 3′ end, and the other allelic probe is complementary to the alternative allele. The two allelic probes compete to anneal to the target DNA adjacent to the common probe and generate double-stranded DNA containing a nick at the allelic site. Only the allelic probe that is perfectly matched to the target DNA is ligated to the common probe by DNA ligase (Figure 5.12).
- *Invasive cleavage* is based on the specificity of recognition of a three-dimensional structure formed when two overlapping oligonucleotides hybridize perfectly to the target DNA (Sobrino et al. 2005). Two oligonucleotide probes, invader and allele specific, anneal to the target DNA with an overlap of one nucleotide. The probe is designed with the allelic base at the overlapping fragment and contains two

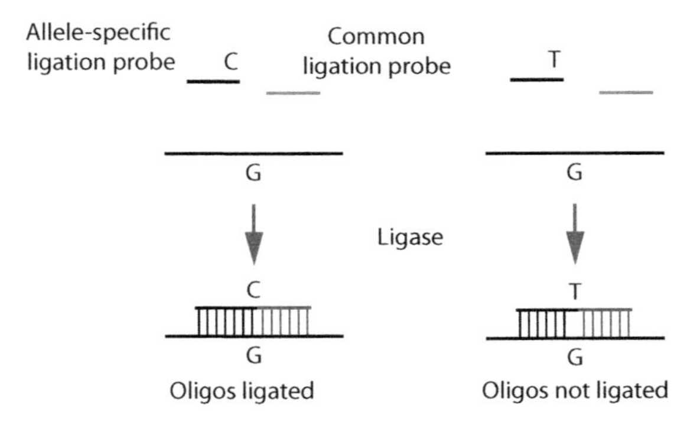

Figure 5.12 Oligonucleotide ligation. Two allele-specific probes and one common ligation probe are required per SNP. The common ligation probe is hybridized adjacent to the allele-specific probe. The ligase joins both allele-specific and common probes when a perfect match of the allele-specific probe is found. (Redrawn from Sobrino, B. et al., *Forensic Sci. Int.*, 154, 181–194, 2005.)

regions. The invader is complementary to the sequence on the 3′ end of the SNP. When the allele-specific probe is complementary to the polymorphic base, it overlaps the 3′ end of the invader to form the structure that is recognized and cleaved by the Flap endonuclease and then releases the 5′ arm of the allele-specific probe (Figure 5.13). If there is a mismatch, the formed structure will not be recognized by

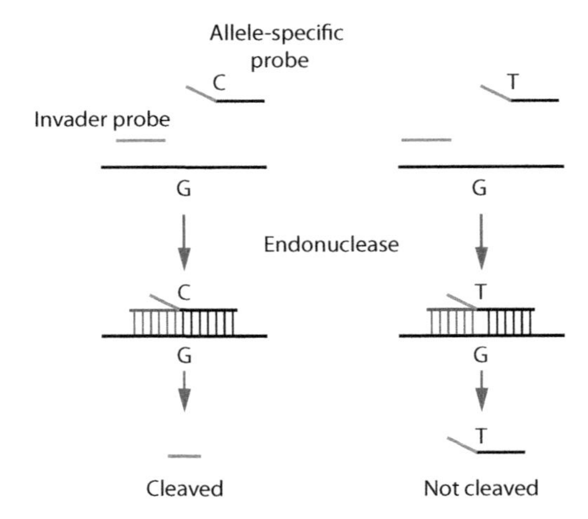

Figure 5.13 Invasive cleavage. The oligonucleotides require an invader probe and allele-specific probes that anneal to the target DNA with an overlap of one nucleotide. When the allele-specific probe is complementary to the polymorphic base, it overlaps the 3′ end of the invader, forming a structure that is recognized and cleaved by the Flap endonuclease, releasing the 5′ arm of the allele-specific probe. (Redrawn from Sobrino, B. et al., *Forensic Sci. Int.*, 154, 181–194, 2005.)

the Flap endonuclease, and the cleavage will not occur. FRET can be applied to this system; a reporter dye is placed on the 5′ arm, and the quencher is placed on the complementary region of the probe. The cleavage event removes the fluorophore and enhances fluorescence.

- *Diversity array technology* (DArT) is a microarray platform that involves fluorescent labeling and hybridization to generate large-scale markers, especially SNPs (Jaccoud et al. 2001). DArT first reduces the complexity of a genome by digestion with two restriction enzymes, rare and frequent cutters, similarly to AFLP. DNA fragments are ligated with adapters and then amplified to form what are regarded as genome representations. The genome representations comprise variable DNA fragments to be hybridized with reference DNA on the microarray. As a result, DArT markers are generated. The DNA reference is prepared from a clone library of the same species and spotted on the microarray.
- *DArTseq* is a newer DArT platform that still keeps the original concept of genome complexity reduction but specifically uses restriction enzymes that can separate low-copy sequences in genomes. DArTseq, instead of using a microarray, deploys an NGS platform to sequence genome representations. The key advantage of DArTseq over the previous DArT is that a much larger number of markers are generated to benefit high-resolution mapping (Sánchez-Sevilla et al. 2015). This technology has been proved in octoploid strawberry, where DArT generated hundreds of SNPs but DArTseq produced almost tens of thousands. In *Capsicum*, 20,000–40,000 SNPs can be identified.

5.3 Molecular mapping of resistance to chili anthracnose

The stage of chili fruit maturity has been shown to play an important role in the differential expression of the resistance genes to *Colletotrichum* spp., the cause of anthracnose (Mahasuk et al. 2009a,b, 2013; Temiyakul et al. 2012). Durable resistance to chili anthracnose, therefore, requires an efficient selection of the resistance genes from different fruit stages. Attempts have been made to locate the anthracnose resistance genes so that MAS can be performed to accurately select the traits based on different stages of fruit maturity.

The very first map of anthracnose resistance in chili was derived from an interspecific population between *C. annuum* and a resistant accession of *C. chinense* (Voorrips et al. 2004). The map was constructed from AFLP markers forming 26 linkage groups (LGs) with 1060 cM coverage. The 26 LGs did not accurately represent the *Capsicum* chromosome number, which was 12, indicating that the map was still incomplete. QTL analysis identified four markers located on different LGs linked to the resistance traits; however, only one marker showed significant effects, which explained approximately 40%–77% of the resistance. More importantly, this study did not take the fruit maturity issue into account.

More recent maps were constructed from chili populations derived from the well-known resistant genotypes *C. chinense* "PBC932" and *C. baccatum* "PBC80" and "PBC81." An interspecific map of *C. annuum* and *C. baccatum* "PBC81" (Lee et al. 2010) was constructed based on AFLP and SSR markers. The map consisted of 13 LGs covering 325 cM. Four QTLs were identified, two of which were on LG12, one on LG9, and one on LGC. The two QTLs on LG12 were associated with resistance to *Co. acutatum* Korean isolate (probably the renamed *Co. scovillei*) at both mature green and ripe fruit stages. The QTL on LG9 was linked to resistance to *Co. truncatum* Thai isolate at both green and ripe fruit stages. The fourth QTL, on LGC, was responsible for resistance to *Co. truncatum* on ripe fruit. The markers flanking the QTL regions were further developed to more robust markers, SCAR and CAPS, to be used in the MAS program (Lee et al. 2011).

The most recent maps were published by Sun et al. (2015) and Mahasuk et al. (2016), whereby two different mapping populations were derived from the same resistant source, "PBC932." One population was derived from an intraspecific *C. baccatum,* with "PBC80" as the resistant parent. The "PBC932" resistance was identified on different chromosomes, P2 (Mahasuk et al. 2016) and P5 (Sun et al. 2015). The map by Sun et al. (2015) comprised 345 SSRs, one InDel, and 35 CAPs, forming 14 LGs, which represented all 12 *Capsicum* chromosomes (P1–P12). A total of five QTLs were identified on P3, P5, P7, P10, and P12, with the most interesting QTL on P5. This P5 QTL explained 30%–60% of the resistance to *Co. scovillei* on both the green and the ripe fruit.

The map constructed by Mahasuk et al. (2016) was based on 214 SNPs forming 12 LGs with 824 cM coverage (Figure 5.14). The 12 LGs represented the *Capsicum* chromosomes. One major QTL, identified on P2, harbored the resistance to *Co. truncatum* on both the mature green and the ripe fruit (Figure 5.15). Interestingly, the linkage of the resistance at both fruit stages correlated with the previous genetic study of resistance using the same population (Mahasuk et al. 2009a).

The different locations of QTLs, identified from "PBC932"-derived resistance by both research groups, could indicate the identification of different resistance genes. The resistance studied by Sun et al. (2015) responded to *Co. scovillei,* while the resistance studied by Mahasuk et al. (2016) responded to *Co. truncatum.* Therefore, different resistance genes were most likely activated by different elicitors (pathogenic factors) based on the gene for gene concept (Flor 1955; Wang & Wang 2018).

To date, the resistance to anthracnose from "PBC80" has been mapped on P4 using an intraspecific population (Mahasuk et al. 2016).

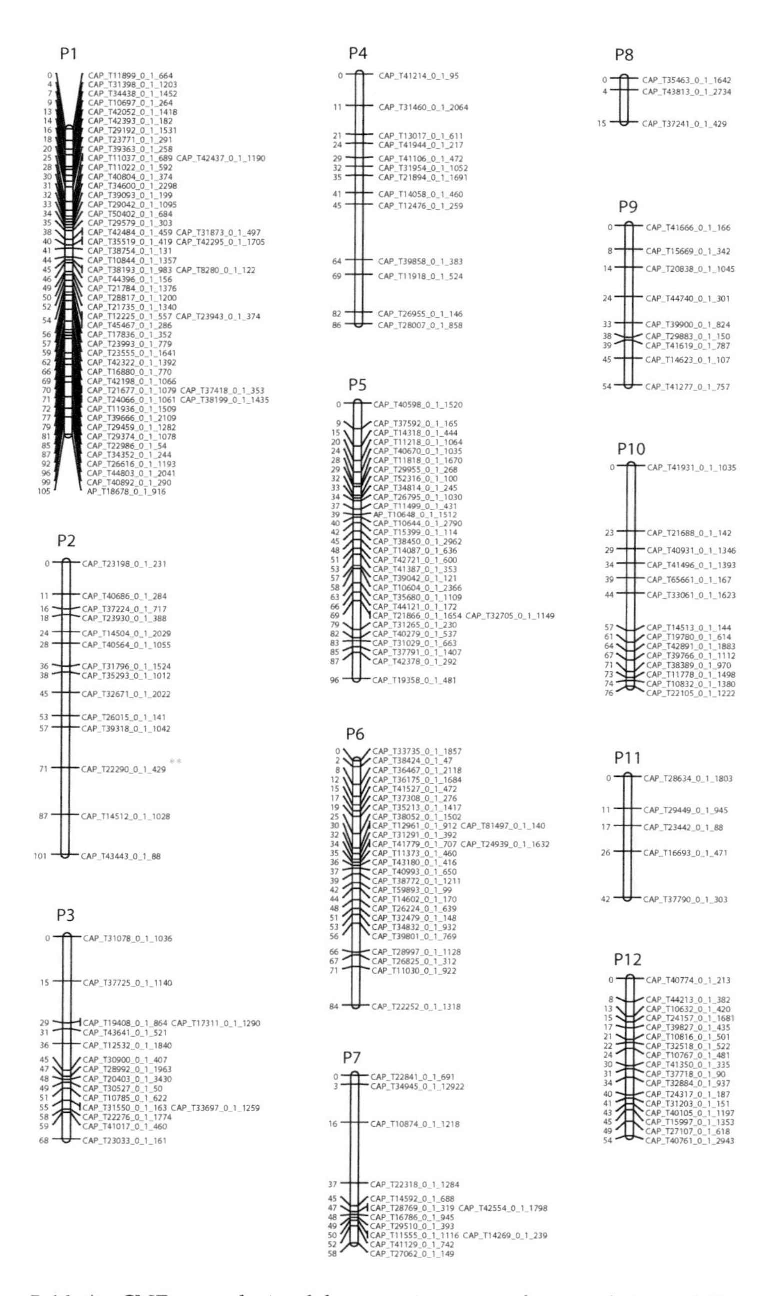

Figure 5.14 An SNP map derived from an interspecific population of *C. annuum* cv. "Bangchang" and *C. chinense* "PBC932." Resistance to *Co. truncatum* "158ci" on mature green and ripe fruit maturity stages is located on chromosome 2 (P2) as indicated with **. (Modified from Mahasuk, P. et al., *Mol. Breed.*, 36, 10, 2016.)

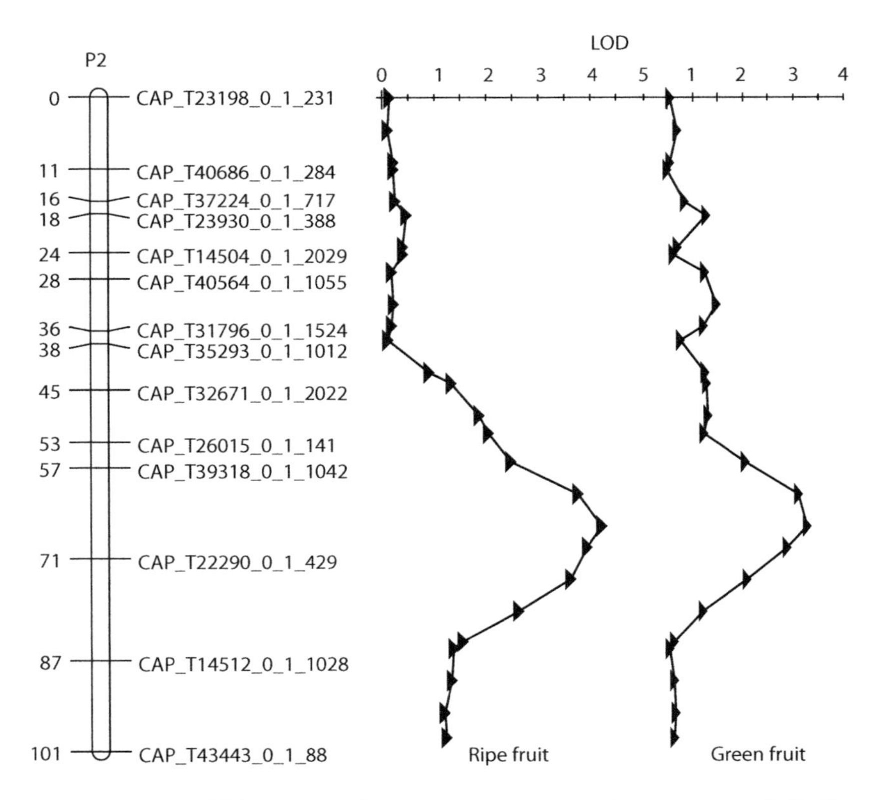

Figure 5.15 Two QTLs responsible for resistance to *Co. truncatum* "158ci" on mature green and ripe fruit are located on the same position of chromosome 2 (P2), based on the map of *C. annuum* cv. "Bangchang" and *C. chinense* "PBC932." (Modified from Mahasuk, P. et al., *Mol. Breed.*, 36, 10, 2016.)

The *C. baccatum* map was constructed based on 403 SNPs that were discovered by *C. baccatum* transcriptome sequencing. The map contained 12 LGs, which represented the *Capsicum* chromosomes with 1,270 cM coverage. QTL analysis with interval mapping identified a total of three QTLs, of which one was major and two were minor QTLs located on P4, P9, and P12. respectively (Figure 5.16). All the QTLs in this map were linked to resistance on the ripe fruit only.

The major QTL on P4 was actually responsible for three resistance traits that were assessed with different *Colletotrichum* pathotypes and inoculation methods. Mahasuk et al. (2016) compared two *Co. scovillei* pathotypes, "MJ5" and "313," which had been classified as Pathotypes 3 and 2, respectively, according to Mongkolporn et al. (2010), and two inoculation methods: microinjection (MI) and high-pressure spray (HP). As a result, three bioassay combinations were performed: MI/MJ5, MI/313,

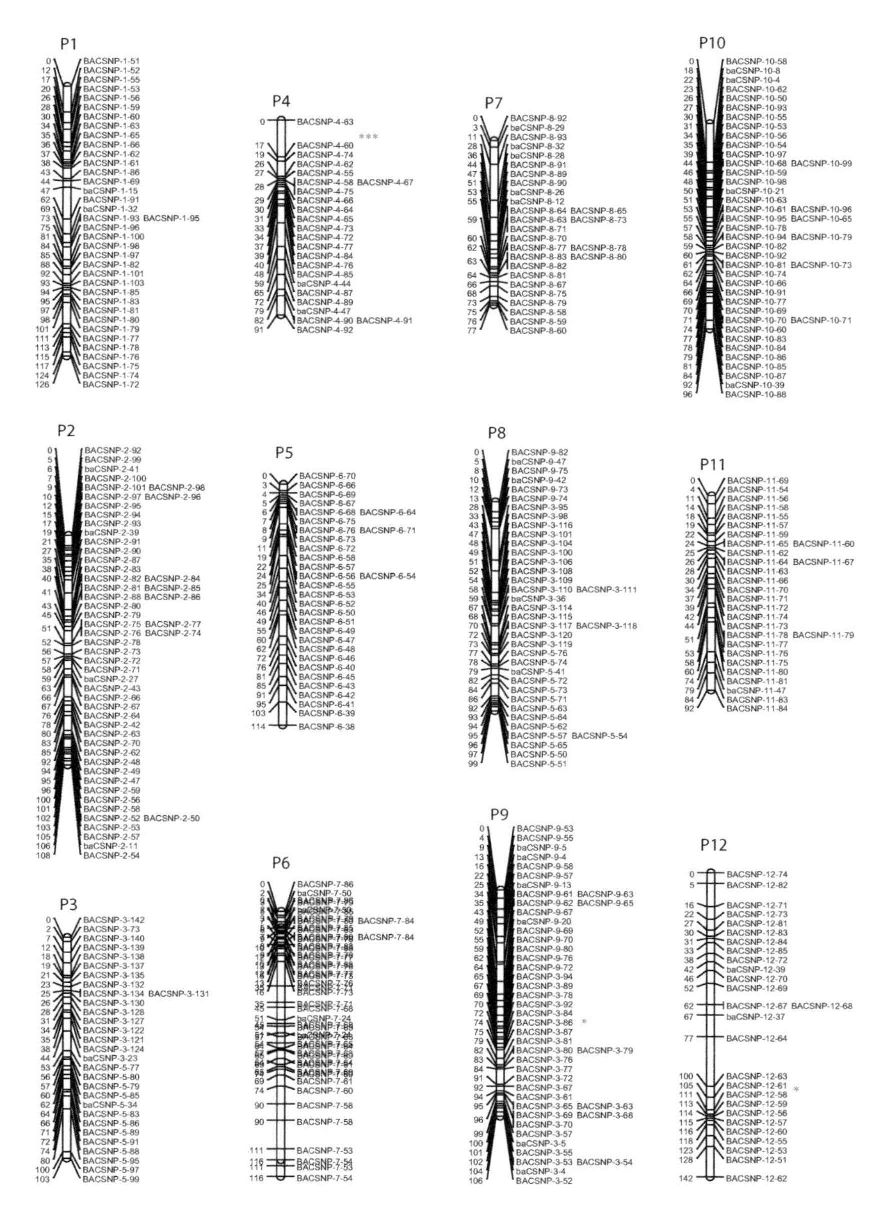

Figure 5.16 An SNP map derived from a *C. baccatum* population crossed between the anthracnose-resistant "PBC80" and the susceptible "CA1316." Three major QTLs responsible for resistance to *Co. scovillei* "MJ5" on ripe fruit are located on chromosome 4 (P4), and two minor QTLs are on P9 and P12, respectively. All the QTL locations are indicated with *. (Modified from Mahasuk, P. et al., *Mol. Breed.*, 36, 10, 2016.)

and HP/MJ5. Three QTLs responsible for the resistance traits assessed by all three bioassay combinations were identified on the same position of P4, two of which are shown in Figure 5.17.

A more recent study based on comparative genome analysis by Kim et al. (2017) has identified 64 nucleotide binding leucine-rich repeat proteins (NLRs) clustered on P3, which corresponded to resistance to *Co. truncatum* (anthracnose) based on the genetic information previously studied in the anthracnose-resistant *C. baccatum* "PBC81." NLRs represent a gene family that mostly is functionally disease resistant. The NLR clustering character was suggested to be a result of retroduplication influenced by LTR retrotransposons, as they are often found to be co-localized. Retroduplication has an impact on the emergence of new disease resistance genes (Kim et al. 2017).

However, the location of the QTL responsible for anthracnose resistance derived from "PBC81" was previously reported to be on P9 (Lee et al. 2011), which differed from the report by Kim et al. (2017). The difference in location of the resistance genes reported by the two studies was possibly due to the chromosomal translocations between P3 and P9 of *C. annuum* and *C. baccatum*, as previously explained in Chapter 2.

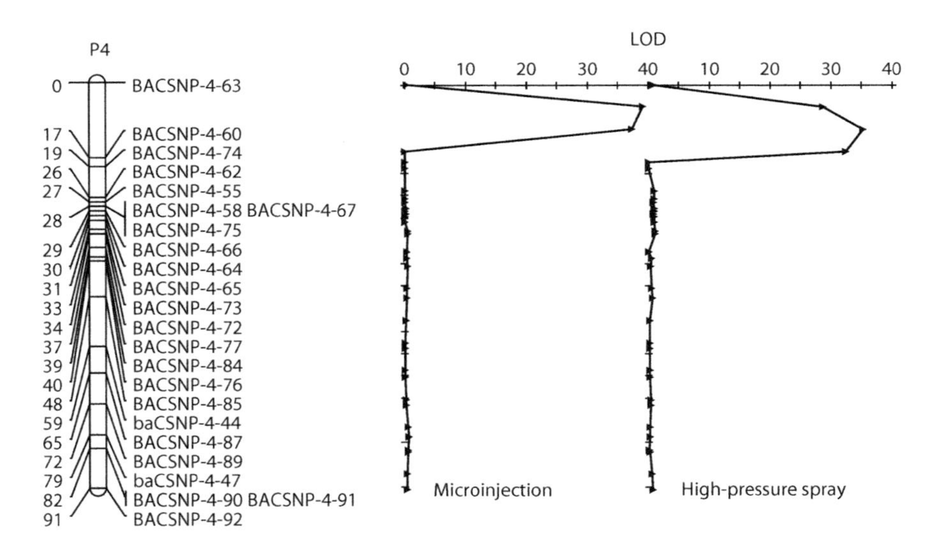

Figure 5.17 Two major QTLs responsible for resistance to *Col. scovillei* "MJ5" on ripe fruit inoculated with two different methods, microinjection and high-pressure spray, are located on the same position of P4, based on the map of *C. annuum* cv. "Bangchang" and *C. chinense* "PBC932." (Modified from Mahasuk, P. et al., *Mol. Breed.*, 36, 10, 2016.)

5.4 *Gene expression in relation to anthracnose resistance*

Differential gene expression in response to chili anthracnose among different fruit maturity stages has been investigated in a series of studies from 1999 to 2014 using one *Capsicum–Colletotrichum* pathosystem. Differential reactions were observed on unripe or green and ripe fruit of a *C. annuum* genotype that was inoculated with *Co. gloeosporioides* isolate "KG13" (Kim et al. 1999). This isolate, "KG13," caused typical anthracnose symptoms on non-wounded and wounded green fruit as well as wounded ripe fruit. In contrast, the non-wounded ripe fruit was not infected; hence, the reaction was incompatible.

An mRNA differential display technique was initially used to identify genes that were expressed differently in healthy and infected green and ripe fruit. Several genes were reported, including thionin, defensin, cytochrome P450, dehydrogenase, P23 protein, NP24 protein, esterase, MADS-box, and thaumatin-like gene (Oh et al. 2003). Some of the identified genes were studied further to understand their roles in defense mechanisms.

Cytochrome P450 was shown to be highly up-regulated in the incompatible reactions in ripe chili fruit but not in the compatible reaction of the green fruit (Oh et al. 1999a). The induction of the gene was also caused by wounding and a jasmonic acid treatment during ripening. The gene was localized in the epidermal cell layers. Cytochrome P450 was suggested to act against the fungal invasion and colonization in the early stage of the infection process.

Thionin (Oh et al. 1999b) was shown to be rapidly induced in ripe fruit after inoculation and reached a peak from 48 hours after inoculation (HAI) until 72 HAI. In the susceptible green fruit, the thionin expression was later and reached its peak at 72 HAI. However, Lee et al. (2000) reported that thionin appeared to be expressed at a higher level and more rapidly in green fruit than in ripe fruit. Thionin is a small, basic, cysteine-rich protein with antimicrobial and toxic properties, which responded to various pathogens, such as *Xanthomonas campestris* (Jung & Hwang 2000), and stress-inducing compounds such as jasmonic acid.

Defensin was initially identified by Oh et al. (1999b) with contrasting activity to thionin. Defensin was suppressed by a *Colletotrichum* infection in ripe fruit. However, it was not until 2014 that defensin was further characterized to show its functional antifungal activity (Seo et al. 2014). Defensin was shown to inhibit 50% of fungal growth. A defensin-transgenic chili was developed and showed the ability to reduce anthracnose lesion formation and fungal colonization.

Thaumatin-like (Kim et al. 2002) and esterase (Ko et al. 2005) genes were isolated from the incompatible ripe chili fruit. Both genes were also induced by wounding and jasmonic acid. The activity of the thaumatin gene was localized in the intercellular spaces among cortical cells, while that of the esterase gene was in the epidermal and cortical cell layers of infected ripe fruit and not in infected green fruit.

Gene expression studies related to HR resistance to chili anthracnose were conducted in *C. baccatum* "PBC80" (Soh et al. 2012; Temiyakul 2012) and *C. chinense* "Bhut Jolokia" (Mishra et al. 2017). Both studies in "PBC80" identified *PR10* as showing rapid up-regulation in response to the *Colletotrichum* infection. PR proteins are a group of small acidic proteins involved in intracellular defense and are normally induced by pathogen infection or stresses. PR10 was found to be highly expressed as early as 6 HAI in both mature green and ripe "PBC80" fruit and highly maintained after inoculation. PR10 clones isolated from "PBC80" and susceptible *C. annuum* genotypes were almost identical; however, *in vitro* enzymatic activity analysis revealed that the recombinant "PBC80" PR10 had higher ribonucleolytic activity and less sensitivity to proteinase than the "susceptible" PR10 (Soh et al. 2012).

A comparative study of defense gene expressions was conducted in two chili genotypes: resistant "Bhut Jolokia" and susceptible "Teja Jhal" (Mishra et al. 2017). A total of 17 defense-related genes were selected for the expression study in response to anthracnose resistance (*Co. truncatum*) of wound-inoculated fruit from three to nine days. Thirteen genes were found to be significantly up-regulated in the resistant reaction of "Bhut Jolokia," six of which were defense signaling genes: defensin 1.2, lipoxygenase 3, allene oxide synthase, phenylalanine ammonia lyase 3, thionin, and 1-aminoacyclopropane 1-carboxylate (ACC) synthase 2. Two were PR genes, PR2 and PR5, and five were defense-responsive transcription factors: *WRKY33*, *CaMYB*, *EREBP1*, *CaNAC*, and *bZIP10*. Although the study covered both the green and ripe fruit stages, the influence of different fruit maturity stages was not considered.

5.5 Conclusions and remarks

Resistance to anthracnose is a complicated trait due to the host–pathogen interaction and complexity, including diverse species of the pathogen, confounded by fruit maturity stages. Conventional selection by fruit bioassay is therefore prone to error, leading to wrong phenotype selection. Also, the selection must be performed on the fruit only; hence, the selection process is tedious and time consuming. Resistance has been identified in species related to *C. annuum*: *C. chinense* "PBC932" and *C. baccatum* "PBC80" and "PBC81." "PBC932" is readily crossed with some *C. annuum* cultivars. QTLs for the "PBC932" resistance were located on chromosomes 2 and 5. Regarding the resistance in *C. baccatum*, only "PBC80"-derived resistance was mapped on chromosome 4. The linked markers from both sources can be helpful in marker-assisted breeding for anthracnose resistance. Since the *Capsicum* genomes have been sequenced, isolation of the resistance genes is likely to happen soon.

References

Adikaram NKB, Brown AE, Swinburne TR. 1982. Phytoalexin involvement in the latent infection of *Capsicum annuum* L. fruit by *Glomerella cingulata* (Stonem.). *Physiological Plant Pathology* 21: 161–170.

Aguilera PM, Debat HJE, Scaldaferro MA, Marti DA, Grabiele M. 2016. FISH-mapping of the 5S rDNA locus in chili peppers (*Capsicum*-Solanaceae). *Annals of the Brazilian Academy of Sciences* 88: 117–125.

Ahmed N, Dey SK, Hundal JS. 1991. Inheritance of resistance to anthracnose in chilli. *Indian Phytopathology* 44: 402–403.

Ahn Y-K, Karna S, Jun T-H, Yang E-Y, Lee H-E, Kim J-H, Kim J-H. 2016. Complete genome sequencing and analysis of *Capsicum annuum* varieties. *Molecular Breeding* 36: 140.

Ahuja I, Kissen R, Bones AM. 2012. Phytoalexins in defense against pathogens. *Trends in Plant Science* 17: 73–90.

Albrecht E, Zhang D, Mays AD, Saftner RA, Stommel JR. 2012. Genetic diversity in *Capsicum baccatum* is significantly influenced by its ecogeographical distribution. *BMC Genetics* 13: 68.

Albrecht E, Zhang D, Saftner RA, Stommel JR. 2011. Genetic diversity and population structure of *Capsicum baccatum* genetic resources. *Genetic Resources and Crop Evolution* 59: 517–538.

Appleby N, Edwards D, Batley J. 2009. New technologies for ultra-high throughput genotyping in plants. In *Plant Genomics: Methods and Protocols,* Vol. 513 (ed. DJ Somers, et al.), pp. 19–39. Humana Press, New York.

Arimboor R, Natarajan RB, Menon KR, Chandrasekhar LP, Moorkoth V. 2015. Red pepper (*Capsicum annuum*) carotenoids as a source of natural food colors: Analysis and stability—a review. *Journal of Food Science and Technology* 52: 1258–1271.

Ashrafi H, Hill T, Stoffel K, Kozik A, Yao J, Chin-Wo SR, Van Deynze A. 2012. *De novo* assembly of the pepper transcriptome (*Capsicum annuum*): A benchmark for in silico discovery of SNPs, SSRs and candidate genes. *BMC Genomics* 13: 571.

Auyong ASM, Ford R, Taylor PWJ. 2012. Genetic transformation of *Colletotrichum truncatum* associated with anthracnose disease of chili by random insertional mutagenesis. *Journal of Basic Microbiology* 52: 372–382.

Auyong ASM, Ford R, Taylor PWJ. 2015. The role of cutinase and its impact on pathogenicity of *Colletotrichum truncatum*. *Journal of Plant Pathology and Microbiology* 6: 259–269.

AVRDC. 1999. *AVRDC Report 1998.* p. 148. Asian Vegetable Research and Development Center, Shanhua, Tainan, Taiwan.

Bailey JA, O'Connell RJ, Pring RJ, Nash C. 1992. Infection strategies of *Colletotrichum* species. In *Colletotrichum: Biology, Pathology and Control* (ed. JA Bailey, MJ Jeger), pp. 88–120. CAB International, Wallingford, UK.

Baral JB, Bosland PW. 2004. Unraveling the species dilemma in Capsicum frutescens and C. chinense (Solanaceae): A multiple evidence approach using morphology, molecular analysis, and sexual compatibility. *Journal of the American Society for Horticultural Science* 129: 826–832.

Barboza GE. 2011. Lectotypifications, synonymy, and a new name in *Capsicum* (Solanoideae, Solanaceae). *PhytoKeys* 2: 23–38.

Barboza GE, Agra MF, Romero MV, Scaldaferro MA, Moscone EA. 2011. New endemic species of *Capsicum* (Solanaceae) from the Brazilian Caatinga: Comparison with the re-circumscribed *C. parvifolium*. *Systematic Botany* 36: 768–781.

Barboza GE, Bianchetti de BL. 2005. Three new species of *Capsicum* (Solanaceae) and a key to the wild species from Brazil. *Systematic Botany* 30: 863–871.

Barroso PA, Rêgo MM, Rêgo ER, Soares WS. 2015. Embryogenesis in the anthers of different ornamental pepper (*Capsicum annuum* L.) genotypes. *Genetics and Molecular Research* 14: 13349–13363.

Behjati S, Tarpey PS. 2013. What is next generation sequencing? *Archives of Disease in Childhood Education and Practice Edition* 98: 236–238.

Berthouly-Salazar C, Mariac C, Couderc M, Pouzadoux J, Floc'h J-B, Vigouroux Y. 2016. Genotyping-by-sequencing SNP identification for crops without a reference genome: Using transcriptome based mapping as an alternative strategy. *Frontiers in Plant Science* 7: 777.

Bock CH, Parker PE, Cook AZ, Gottwald TR. 2008. Visual rating and the use of image analysis for assessing different symptoms of citrus canker on grapefruit leaves. *Plant Disease* 92: 530–541.

Bock CH, Poole GH, Parker PE, Gottwald TR. 2010. Plant disease severity estimated visually, by digital photography and image analysis, and by hyperspectral imaging. *Critical Reviews in Plant Sciences* 29: 59–107.

Bosland PW, Baral JB. 2007. "Bhut Jolokia"—The world's hottest known Chile pepper is a putative naturally occurring interspecific hybrid. *HortScience* 42: 222–224.

Bosland PW, Votava EJ. 2012. *Peppers: Vegetable and Spice Capsicums*, 2nd edition. CAB International, Reading, UK.

Botstein D, White RL, Skolnick M, Davis RW. 1980. Construction of a genetic linkage map in man using restriction fragment length polymorphisms. *American Journal of Human Genetics* 32: 314–331.

Brown AHD. 1989. Core collections: A practical approach to genetic resources management. *Genome* 31: 818–824.

Caicedo AL, Purugganan MD. 2005. Comparative plant genomics: Frontiers and prospects. *Plant Physiology* 138: 545.

Caires NP, Pinho DB, Souza JSC, Silva MA, Lisboa DO, Pereira OL, Furtado GQ. 2014. First report of anthracnose on pepper fruit caused by *Colletotrichum scovillei* in Brazil. *Plant Disease* 98: 1437.

Cannon PF, Damm U, Johnston PR, Weir BS. 2012. *Colletotrichum*—current status and future directions. *Studies in Mycology* 73: 181–213.

Carvalho SIC, Bianchetti LB, Ragassi CF, Ribeiro CSC, Reifschneider FJB, Buso GSC, Faleiro FG. 2017. Genetic variability of a Brazilian *Capsicum frutescens* germplasm collection using morphological characteristics and SSR markers. *Genetics and Molecular Research* 16: gmr16039689.

Carvalho SIC, Ragassi CF, Bianchetti LB, Reifschneider FJB, Buso GSC, Faleiro FG. 2014. Morphological and genetic relationships between wild and domesticated forms of peppers (*Capsicum frutescens* L. and *C. chinense* Jacquin). *Genetics and Molecular Research* 13: 7447–7464.

Coll NS, Epple P, Dangl JL. 2011. Programmed cell death in the plant immune system. *Cell Death and Differentiation* 18: 1247–1256.

Collard BCY, Jahufer ZZ, Brouwer JB, Pang ECK. 2005. An introduction to markers, quantitative trait loci (QTL) mapping and marker-assisted selection for crop improvement: The basic concepts. *Euphytica* 142: 169–196.

Cruz MVR, Urdampilleta JD, Martins ERF, Moscone EA. 2017. Cytogenetic markers for the characterization of *Capsicum annuum* L. cultivars. *Plant Biosystems* 151: 84–91.

Damm U, Cannon PF, Woudenberg JHC, Crous PW. 2012. The *Colletotrichum acutatum* species complex. *Studies in Mycology* 73: 37–113.

Damm U, Woudenberg JHC, Cannon PF, Crous PW. 2009. *Colletotrichum* species with curved conidia from herbaceous hosts. *Fungal Diversity* 39: 45–87.

De Almeida LB, Matos KS, Assis LAG, Hanada RE, da Silva GF. 2017. First report of anthracnose of *Capsicum chinense* in Brazil caused by *Colletotrichum brevisporum*. *Plant Disease*. 101: 1035.

De Oliveira CVS, Matos KS, de Albuquerque DMC, Hanada RE, da Silva GF. 2017. Identification of *Colletotrichum* isolates from *Capsicum chinense* in Amazon. *Genetics and Molecular Research* 16: gmr16029601.

De Silva DD, Ades PK, Crous PW, Taylor PWJ. 2017a. *Colletotrichum* species associated with chili anthracnose in Australia. *Plant Pathology* 66: 254–267.

De Silva DD, Crous PW, Ades PK, Hyde KD, Taylor PWJ. 2017b. Life styles of *Colletotrichum* species and implications for plant biosecurity. *Fungal Biology Reviews* 31: 155–168.

Dean R, van Kan JAL, Pretorius ZA, Hammond-Kosack KE, di Pietro A, Spanu PD, Rudd JJ et al. 2012. The top 10 fungal pathogens in molecular plant pathology. *Molecular Plant Pathology* 13: 414–430.

Diao Y-Z, Zhang C, Liu F, Wang W-Z, Liu L, Cai L, Liu X-L. 2017. *Colletotrichum* species causing anthracnose disease of chili in China. *Persoonia* 38: 20–37.

Didenko VV. 2001. DNA probes using fluorescence resonance energy transfer (FRET): Designs and applications. *BioTechniques* 31: 1106–1121.

Edwards D, Forster JW, Chagne D, Batley J. 2007. What Are SNPs? In *Association Mapping in Plants* (ed. NC Oraguzie, et al.), pp. 41–52. Springer, Berlin.

Eshbaugh WH. 2012. The taxonomy of the genus *Capsicum*. In *Peppers: Botany, Production and Uses* (ed. VM Russo, et al.), pp. 14–28. CAB International, Wallingford, UK.

FAO. 2017a. Crops. www.fao.org/faostat/en/#data/QC (Date accessed 15 October 2017).

FAO. 2017b. Value of agricultural production. www.fao.org/faostat/en/#data/QV (Date accessed 15 October 2017).

Ferreira A, da Silva MF, da Costa e Silva L, Cruz CD. 2006. Estimating the effects of population size and type on the accuracy of genetic maps. *Genetics and Molecular Biology* 29: 187–192.

Flor HH. 1955. Host-parasite interaction in flax rust—its genetics and other implications. *Phytopathology* 45: 680–685.

Francl LJ. 2001. The disease triangle: A plant pathological paradigm revisited. *The Plant Health Instructor*. doi:10.1094/PHI-T-2001-0517-01.

Frankel OH. 1984. Genetic perspectives of germplasm conservation. In *Genetic Manipulation: Impact on Man and Society* (ed. WK Arber, et al.), pp. 161–170. Cambridge University Press, Cambridge, UK.

Frankel OH, Brown AHD. 1984. Plant genetic resources today: A critical appraisal. In *Crop Genetic Resources: Conservation and Evaluation* (ed. JHW Holden, JT Williams), pp. 249–257. George Allen and Unwin, London, UK.

Fu Y-B. 2015. Understanding crop genetic diversity under modern plant breeding. *Theoretical and Applied Genetics* 128: 2131–2142.

Gan P, Narusaka M, Kumakura N, Tsushima A, Takano Y, Narusaka Y, Shirasu K. 2016. Genus-wide comparative genome analyses of *Colletotrichum* species reveal specific gene family losses and gains during adaptation to specific infection llfestyles. *Genome Biology and Evolution* 8: 1467–1481.

Ganal MW, Altmann T, Röder MS. 2009. SNP identification in crop plants. *Current Opinion in Plant Biology* 12: 211–217.

George L, Narayanaswamy S. 1973. Haploid *Capsicum* through experimental androgenesis. *Protoplasma* 78: 467–470.

Giovannoni JJ, Wing RA, Ganal MW, Tanksley SD. 1991. Isolation of molecular markers from specific chromosomal intervals using DNA pools from existing mapping populations. *Nucleic Acids Research* 19: 6553–6558.

González-Pérez S, Garcés-Claver A, Mallor C, Sáenz de Miera LE, Fayos O, Pomar F, Merino F, et al. 2014. New insights into *Capsicum* spp. relatedness and the diversification process of *Capsicum annuum* in Spain. *PLoS ONE* 9: e116276.

Gottwald TR, Timmer LW, McGuire RG. 1989. Analysis of disease progress of citrus canker in nurseries in Argentina. *Phytopathology* 79: 1276–1283.

Gudeva LK, Mitrev S, Maksimova V, Spasov D. 2013. Content of capsaicin extracted from hot pepper (*Capsicum annuum* ssp. *microcarpum* L.) and its use as an ecopesticide. *Hemijska Industrija* 67: 671–675.

Han J-H, Chon J-K, Ahn J-H, Choi I-Y, Lee Y-H, Kim KS. 2016a. Whole genome sequence and genome annotation of *Colletotrichum acutatum*, causal agent of anthracnose in pepper plants in South Korea. *Genomics Data* 8: 45–46.

Han K, Jeong H-J, Yang H-B, Kang S-M, Kwon J-K, Kim S, Choi D et al. 2016b. An ultra-high-density bin map facilitates high-throughput QTL mapping of horticultural traits in pepper (*Capsicum annuum*). *DNA Research* 23: 81–91.

Hardison RC. 2003. Comparative genomics. *PLoS Biology* 1: e58.

Hartl DL, Jones EW. 2005. *Genetics: Analysis of Genes and Genomes*, 6th edition. Jones & Bartlett Publishers, Sudbury, MA.

Hayward AC, Tollenaere R, Dalton-Morgan J, Batley J. 2015. *Molecular Marker Applications in Plants*. Springer, Science and Business Media, New York.

Heiser CBJ, Pickersgill B. 1975. Names for the bird peppers (*Capsicum*-Solanaceae). *Baileya* 19: 151–156.

Hill T, Ashrafi H, Chin-Wo SR, Stoffel K, Truco M-J, Kozik A, Michelmore R et al. 2015. Ultra-high density, transcript-based genetic maps of pepper define recombination in the genome and synteny among related species. *G3: Genes|Genomes|Genetics* 5: 2341–2355.

Hong JK, Hwang BK. 1998. Influence of inoculum density, wetness duration, plant age, inoculation method, and cultivar resistance on infection of pepper plants by *Colletotrichum coccodes*. *Plant Disease* 82: 1079–1083.

Hunziker AT. 2001. *Genera Solanacearum: The Genera of Solanaceae Illustrated, Arranged According to a New System*. Gantner Verlag, Ruggell, Leichtenstein, Germany.

Hyde KD, Cai L, Cannon PF, Crouch JA, Crous PW, Damm U, Goodwin PH et al. 2009a. *Colletotrichum* – names in current use. *Fungal Diversity* 39: 147–182.

Hyde KD, Cai L, McKenzie EHC, Yang YL, Zhang JZ, Prihastuti H. 2009b. *Colletotrichum*: A catalogue of confusion. *Fungal Diversity* 39: 1–17.

Ibiza VP, Blanca J, Cañizares J, Nuez F. 2012. Taxonomy and genetic diversity of domesticated *Capsicum* species in the Andean region. *Genetic Resources and Crop Evolution* 59: 1077–1088.

IBPGR. 1983. *Genetic Resources of Capsicum: A Global Plan of Action*. International Board for Plant Genetic Resources, Rome, Italy.

IPNI. 2012. The International Plant Names Index. www.ipni.org (Date accessed 24 September 2017).

Irikova T, Grozeva S, Rodeva V. 2011. Anther culture in pepper (*Capsicum annuum* L.) in vitro. *Acta Physiologiae Plantarum* 33: 1559–1570.

Jaccoud D, Peng K, Feinstein D, Killian A. 2001. Diversity arrays: A solid state technology for sequence information independent genotyping. *Nucleic Acids Research* 29: 1–7.

Jansen RC, Stam P. 1994. High resolution of quantitative traits into multiple loci via interval mapping. *Genetics* 136: 1447–1455.

Jha TB, Dafadar A, Ghorai A. 2012. New genetic resource in *Capsicum* L. from Eastern Himalayas. *Plant Genetic Resources: Characterization and Utilization* 10: 141–144.

Jiang G-L. 2013. Molecular Markers and Marker-Assisted Breeding in Plants. In *Plant Breeding from Laboratories to Fields*, doi:10.5772/52583. (ed. SB Andersen), pp. 45-83. InTech, Rijeka.

Jo YD, Choi Y, Kim D-H, Kim B-D, Kang B-C. 2014. Extensive structural variations between mitochondrial genomes of CMS and normal peppers (*Capsicum annuum* L.) revealed by complete nucleotide sequencing. *BMC Genomics* 15: 561.

Jones JDG, Dangl JL. 2006. The plant immune system. *Nature* 444: 323–329.

Jones N, Ougham H, Thomas H, Pašakinskienė I. 2009. Markers and mapping revisited: Finding your gene. *New Phytologist* 183: 935–966.

Jung HW, Hwang BK. 2000. Isolation, partial sequencing, and expression of pathogenesis-related cDNA genes from pepper leaves infected by *Xanthomonas campestris* pv. *vesicatoria*. *Molecular Plant-Microbe Interactions* 13: 136–142.

Kanchana-udomkan C, Taylor PWJ, Mongkolporn O. 2004. Development of a bioassay to study anthracnose infection of chili fruit caused by *Colletotrichum capsici*. *Thai Journal of Agricultural Science* 37: 293–297.

Kang J-H, Yang H-B, Jeong H-S, Choe P, Kwon J-K, Kang B-C. 2014. Single nucleotide polymorphism marker discovery from transcriptome sequencing for marker-assisted backcrossing in *Capsicum*. *Korean Journal of Horticultural Science and Technology* 32: 535–543.

Kantar MB, Anderson JE, Lucht SA, Mercer K, Bernau V, Case KA, Le NC et al. 2016. Vitamin variation in *Capsicum* Spp. provides opportunities to improve nutritional value of human diets. *PLoS ONE* 11: e0161464.

Kanto T, Uematsu S, Tsukamoto T, Moriwaki J, Yamagishi N, Usami T, Sato T. 2014. Anthracnose of sweet pepper caused by *Colletotrichum scovillei* in Japan. *Journal of General Plant Pathology* 80: 73–78.

Katoch A, Sharma P, Sharma PN. 2017. Identification of *Colletotrichum* spp. associated with fruit rot of *Capsicum annuum* in North Western Himalayan region of India using fungal DNA barcode markers. *Journal of Plant Biochemistry and Biotechnology* 26: 216–223.

Kesawat MS, Das BK. 2009. Molecular markers: Its application in crop improvement. *Journal of Crop Science and Biotechnology* 12: 169–181.

Kim KD, Oh BJ, Yang J. 1999. Differential interactions of a *Colletotrichum gloeosporioides* isolate with green and red pepper fruits. *Phytoparasitica* 27: 97–106.

Kim K-H, Yoon J-B, Park H-G, Park EW, Kim YH. 2004. Structural modifications and programmed cell death of chili pepper fruit related to resistance responses to *Colletotrichum gloeosporioides* infection. *Phytopathology* 94: 1295–1304.

Kim J-S, Jee H-J, Gwag J-G, Kim C-K, Shim C-K. 2010. Evaluation on red pepper germplasm lines (*Capsicum* spp.) for resistance to anthracnose caused by *Colletotrichum acutatum*. *Plant Pathology Journal* 26: 273–279.

Kim KW, Chung HK, Cho GT, Ma KH, Chandrabalan D, Gwag JG, Kim TS et al. 2007a. PowerCore: A program applying the advanced M strategy with a heuristic search for establishing allele mining sets. *Bioinformatics* 23: 2155–2162.

Kim S, Park J, Yeom S-I, Kim Y-M, Seo E, Kim K-T, Kim M-S et al. 2017. New reference genome sequences of hot pepper reveal the massive evolution of plant disease-resistance genes by retroduplication. *Genome Biology* 18: 210.

Kim S, Park M, Yeom S-I, Kim Y-M, Lee JM, Lee H-A, Seo E et al. 2014. Genome sequence of the hot pepper provides insights into the evolution of pungency in *Capsicum* species. *Nature Genetics* 46: 270–278.

Kim SH, Yoon JB, Do JW, Park HG. 2007b. Resistance to anthracnose caused by *Colletotrichum acutatum* in chili pepper (*Capsicum annuum* L.). *Journal of Crop Science and Biotechnology* 10: 277–280.

Kim SH, Yoon JB, Do JW, Park HG. 2008a. A major recessive gene associated with anthracnose resistance to *Colletotrichum capsici* in chili pepper (*Capsicum annuum* L.). *Breeding Science* 58: 137–141.

Kim SH, Yoon JB, Park HG. 2008b. Inheritance of anthracnose resistance in a new genetic resource, *Capsicum baccatum* PI594137. *Journal of Crop Science and Biotechnology* 11: 13–16.

Kim T-S, Lee J-R, Raveendar S, Lee G-A, Jeon Y-A, Lee H-S, Ma K-H et al. 2016. Complete chloroplast genome sequence of *Capsicum baccatum* var. *baccatum*. *Molecular Breeding* 36: 110.

Kim YS, Lee HH, Ko MK, Song CE, Bae C-Y, Lee YH, Oh B-J. 2001. Inhibition of fungal appressorium formation by pepper (*Capsicum annuum*) esterase. *Molecular Plant-Microbe Interactions* 14: 80–85.

Kim YS, Park JY, Kim KS, Ko MK, Cheong SJ, Oh BJ. 2002. A thaumatin-like gene in nonclimacteric pepper fruits used as molecular marker in probing disease resistance, ripening, and sugar accumulation. *Plant Molecular Biology* 49: 125–135.

Ko MK, Jeon WB, Kim KS, Lee HH, Seo HH, Kim YS, Oh B-J. 2005. A *Colletotrichum gloeosporioides*-induced esterase gene of nonclimacteric pepper (*Capsicum annuum*) fruit during ripening plays a role in resistance against fungal infection. *Plant Molecular Biology* 58: 529–541.

Konieczny A, Ausubel FM. 1993. A procedure for mapping Arabidopsis mutations using co-dominant ecotype-specific PCR-based markers. *The Plant Journal* 4: 403–410.

Kulkarni SS, Bodake UM, Pathade GR. 2011. Extraction of natural dye from chili (*Capsicum annuum*) for textile coloration. *Universal Journal of Environmental Research and Technology* 1: 58–63.

Lee J, Do J, Yoon J. 2011. Development of STS markers linked to the major QTLs for resistance to the pepper anthracnose caused by *Colletotrichum acutatum* and *C. capsici*. *Horticulture, Environment, and Biotechnology* 52: 596–601.

Lee J, Hong J-H, Do JW, Yoon JB. 2010. Identification of QTLs for resistance to anthracnose to two *Colletotrichum* species in pepper. *Journal of Crop Science and Biotechnology* 13: 227–233.

Lee JM, Nahm SH, Kim YM, Kim BD. 2004. Characterization and molecular genetic mapping of microsatellite loci in pepper. *Theoretical and Applied Genetics* 108: 619–627.

Lee SC, Lee YK, Kim KD, Hwang BK. 2000. In situ hybridization study of organ- and pathogen-dependent expression of a novel thionin gene in pepper (*Capsicum annuum*). *Physiologia Plantarum* 110: 384–392.

Liao C-Y, Chen M-Y, Chen Y-K, Wang T-C, Sheu Z-M, Kuo K-C, Chang P-FL et al. 2012a. Characterization of three *Colletotrichum acutatum* isolates from *Capsicum* spp. *European Journal of Plant Pathology* 133: 599–608.

Liao C-Y, Chen M-Y, Chen Y-K, Kuo K-C, Chung K-R, Lee M-H. 2012b. Formation of highly branched hyphae by *Colletotrichum acutatum* within the fruit cuticles of *Capsicum* spp. *Plant Pathology* 61: 262–270.

Lin Q, Kanchana-udomkan C, Jaunet T, Mongkolporn O. 2002. Genetic analysis of resistance to pepper anthracnose caused by *Colletotrichum capsici*. *Thai Journal of Agricultural Science* 35: 259–264.

Lin SW, Chou YY, Shieh HC, Ebert AW, Kumar S, Mavlyanova R, Rouamba A et al. 2013. Pepper (*Capsicum* spp.) germplasm dissemination by AVRDC—The World Vegetable Center: An overview and introspection. *Chronica Horticulturae* 53: 21–27.

Lin SW, Gniffke PA, Wang TC. 2007. Inheritance of resistance to pepper anthracnose caused by *Colletotrichum acutatum*. *Acta Horticulturae (ISHS)* 760: 329–334.

Liu F, Cai L, Crous PW, Damm U. 2013. Circumscription of the anthracnose pathogens *Colletotrichum lindemuthianum* and *C. nigrum*. *Mycologia* 105: 844–860.

Liu F, Tang G, Zheng X, Li Y, Sun X, Qi X, Zhou Y et al. 2016. Molecular and phenotypic characterization of Colletotrichum species associated with anthracnose disease in peppers from Sichuan Province, China. *Scientific Reports* 6: 32761.

Mahasuk P, Chinthaisong J, Mongkolporn O. 2013. Differential resistances to anthracnose in *Capsicum baccatum* as responding to two *Colletotrichum* pathotypes and inoculation methods. *Breeding Science* 63: 333–338.

Mahasuk P, Khumpeng N, Wasee S, Taylor PWJ, Mongkolporn O. 2009a. Inheritance of resistance to anthracnose (*Colletotrichum capsici*) at seedling and fruiting stages in chili pepper (*Capsicum* spp.). *Plant Breeding* 128: 701–706.

Mahasuk P, Struss D, Mongkolporn O. 2016. QTLs for resistance to anthracnose identified in two *Capsicum* sources. *Molecular Breeding* 36: 10.

Mahasuk P, Taylor PWJ, Mongkolporn O. 2009b. Identification of two new genes conferring resistance to *Colletotrichum acutatum* in *Capsicum baccatum* L. *Phytopathology* 99: 1100–1104.

Maksimova V, Koleva LG, Ruskovska T, Cvetanovska A, Gulaboski R. 2014. Antioxidative effect of *Capsicum* oleoresins compared with pure capsaicin. *IOSR Journal of Pharmacy* 4: 44–48.

Manzur JP, Fita A, Prohens J, Rodríguez-Burruezo A. 2015. Successful wide hybridization and introgression breeding in a diverse set of common peppers (*Capsicum annuum*) using different cultivated ají (*C. baccatum*) accessions as donor parents. *PLoS ONE* 10: e0144142.

Marin-Felix Y, Groenewald JZ, Cai L, Chen Q, Marincowitz S, Barnes I, Bensch K et al. 2017. Genera of phytopathogenic fungi: GOPHY 1. *Studies in Mycology* 86: 99–216.

Martins KC, Pereira TNS, Souza SAM, Rodrigues R, do Amaral ATJ. 2015. Crossability and evaluation of incompatibility barriers in crosses between *Capsicum* species. *Crop Breeding and Applied Biotechnology* 15: 139–145.

McCarty MF, DiNicolantonio JJ, O'Keefe JH. 2015. Capsaicin may have important potential for promoting vascular and metabolic health. *Open Heart* 2: e000262.

Meekul P, Chunwongse J. 2010. Production and verification of double haploid lines derived from F1 hybrid pepper cvs. "83-168" and "PEPAC25" by anther culture. *Agricultural Science Journal (Thailand)* 41: 203–212.

Metzker ML. 2005. Emerging technologies in DNA sequencing. *Genome Research* 15: 1767–1776.

Metzker ML. 2010. Sequencing technologies—the next generation. *Nature Reviews Genetics* 11: 31–46.

Michelmore RW, Paran I, Kesseli RV. 1991. Identification of markers linked to disease-resistance genes by bulked segregant analysis: A rapid method to detect markers in specific genomic regions by using segregating populations. *Proceedings of the National Academy of Sciences of the United States of America* 88: 9828–9832.

Mills PR, Hodson A, Brown AE. 1992. Molecular differentiation of *Colletotrichum gloeosporioides* isolates infecting tropical fruits. In Colletotrichum*: Biology, Pathology and Control* (ed. JA Bailey, MJ Jeger), pp. 269–288. CABI, Wallingford, UK.

Mishra R, Nanda S, Rout E, Chand SK, Mohanty JN, Joshi RK. 2017. Differential expression of defense-related genes in chilli pepper infected with anthracnose pathogen *Colletotrichum truncatum*. *Physiological and Molecular Plant Pathology* 97: 1–10.

Mongkolporn O, Hanyong S, Chunwongse J, Wasee S. 2015. Establishment of a core collection of chilli germplasm using microsatellite analysis. *Plant Genetic Resources: Characterization and Utilization* 13: 104–110.

Mongkolporn O, Montri P, Supakaew T, Taylor PWJ. 2010. Differential reactions on mature green and ripe chili fruit infected by three *Colletotrichum* species. *Plant Disease* 94: 306–310.

Mongkolporn O, Taylor PWJ. 2011. *Capsicum*. In *Wild Crop Relatives: Genomic and Breeding Resources*, Vol 5: Vegetables (ed. C Kole), pp. 43–57. Springer, New York.

Mongkolporn O, Taylor PWJ. 2018. Chili anthracnose: *Colletotrichum* taxonomy and pathogenicity. *Plant Pathology* 67: in press, doi: 10.1111/ppa.12850

Montri P, Taylor PWJ, Mongkolporn O. 2009. Pathotypes of *Colletotrichum capsici*, the causal agent of chili anthracnose, in Thailand. *Plant Disease* 93: 17–20.

Moscone EA, Baranyi M, Ebert I, Greilhuber J, Ehrendorfer F, Hunziker AT. 2003. Analysis of nuclear DNA content in *Capsicum* (Solanaceae) by flow cytometry and Feulgen densitometry. *Annals of Botany* 92: 21–29.

Moscone EA, Scaldaferro MA, Grabiele M, Cecchini NM, Sánchez García Y, Jarret R, Daviña JR et al. 2007. The evolution of chili peppers (*Capsicum*—Solanaceae): A cytogenetic perspective. *Acta Horticulturae (ISHS)* 745: 137–170.

Moscone EA, Scaldaferro MA, Grabiele M, Romero MV, Debat H, Seijo JG, Acosta MC et al. 2011. Genomic characterization of the chili peppers (*Capsicum*,

Solanaceae) germplasm by classical and molecular cytogenetics. In *Physical Mapping Technologies for the Identification and Characterization of Mutated Genes Contributing to Crop Quality*, pp. 97–104. International Atomic Energy Agency, Vienna, Austria.

Moses M, Umaharan P. 2012. Genetic structure and phylogenetic relationships of *Capsicum chinense*. *Journal of the American Society for Horticultural Science* 137: 250–262.

Muthamilarasan M, Prasad M. 2013. Plant innate immunity: An updated insight into defense mechanism. *Journal of Biosciences* 38: 433–449.

Mutka AM, Bart RS. 2015. Image-based phenotyping of plant disease symptoms. *Frontiers in Plant Science* 5: 734.

Naranjo CA, Poggio L, Brandham PE. 1983. A practical method of chromosome classification on the basis of centromere position. *Genetica* 62: 51–53.

Nasehi A, Kadir J, Rashid TS, Awla HK, Golkhandan E, Mahmodi F. 2016. Occurrence of anthracnose fruit rot caused by *Colletotrichum nymphaeae* on pepper (*Capsicum annuum*) in Malaysia. *Plant Disease* 100: 1244.

Nicolai M, Pisani C, Bouchet JP, Vuylsteke M, Palloix A. 2012. Discovery of a large set of SNP and SSR genetic markers by high-throughput sequencing of pepper (*Capsicum annuum*). *Genetics and Molecular Research* 11: 2295–2300.

Nimmakayala P, Abburi VL, Saminathan T, Alaparthi SB, Almeida A, Davenport B, Nadimi M et al. 2016a. Genome-wide diversity and association mapping for capsaicinoids and fruit weight in *Capsicum annuum* L. *Scientific Reports* 6: 38081.

Nimmakayala P, Abburi V, Saminathan T, Almeida A, Davenport B, Davidson J, Reddy CVCM et al. 2016b. Genome-wide divergence and linkage disequilibrium analyses for *Capsicum baccatum* revealed by genome-anchored single nucleotide polymorphisms. *Frontiers in Plant Science* 7: 1–12.

Novak F. 1974. *Capsicum* haploids. *Zeitschrift fur Pflanzenzuchtung* 72: 46–54.

Nutter FWJ, Teng PS, Shokes FM. 1991. Disease assessment terms and concepts. *Plant Disease* 75: 1187–1188.

O'Connell RJ, Thon MR, Hacquard S, Amyotte SG, Kleemann J, Torres MF, Damm U et al. 2012. Lifestyle transitions in plant pathogenic *Colletotrichum* fungi deciphered by genome and transcriptome analyses. *Nature Genetics* 44: 1060–1067.

Oh BJ, Ko MK, Kim YS, Kim KS, Kostenyuk I, Kee HK. 1999a. A cytochrome P450 gene is differentially expressed in compatible and incompatible interactions between pepper (*Capsicum annuum*) and the anthracnose fungus, *Colletotrichum gloeosporioides*. *Molecular Plant-Microbe Interactions* 12: 1044–1052.

Oh B-J, Ko M-K, Kim KS, Kim YS, Lee HH, Jeon WB, Im KH. 2003. Isolation of defense-related genes differentially expressed in the resistance interaction between pepper fruits and the anthracnose fungus *Colletotrichum gloeosporioides*. *Molecules and Cells* 15: 349–355.

Oh BJ, Ko MK, Kostenyuk I, Shin B, Kim KS. 1999b. Coexpression of a defensin gene and a thionin-like gene via different signal transduction pathways in pepper and *Colletotrichum gloeosporioides* interactions. *Plant Molecular Biology* 41: 313–319.

Oh J, Giallongo F, Frederick T, Pate J, Walusimbi S, Elias RJ, Wall EH et al. 2015. Effects of dietary *Capsicum* oleoresin on productivity and immune responses in lactating dairy cows. *Journal of Dairy Science* 98: 6327–6339.

Olson M, Hood L, Cantor C, Botstein D. 1989. A common language for physical mapping of the human genome. *Science* 245: 1434–1435.

O'Neill J, Brock C, Olesen AE, Andresen T, Nilsson M, Dickenson AH. 2012. Unravelling the mystery of capsaicin: A tool to understand and treat pain. *Pharmacological Reviews* 64: 939–971.

Oo MM, Lim G, Jang HA, Oh S-K. 2017. Characterization and pathogenicity of new record of anthracnose on various chili varieties caused by *Colletotrichum scovillei* in Korea. *Mycobiology* 45: 184–191.

Orita M, Iwahana H, Kanazwa H, Hayashi K, Sekiya T. 1989. Detection of polymorphisms of human DNA by gel electrophoresis as single-strand conformation polymorphisms. *Proceedings of the National Academy of Sciences of the United States of America* 86: 2766–2770.

Pakdeevaraporn P, Wasee S, Taylor PWJ, Mongkolporn O. 2005. Inheritance of resistance to anthracnose caused by *Colletotrichum capsici* in *Capsicum*. *Plant Breeding* 124: 206–208.

Paran I, Michelmore RW. 1993. Development of reliable PCR-based markers linked to downy mildew resistance genes in lettuce. *Theoretical and Applied Genetics* 85: 985–993.

Park HK, Kim BS, Lee WS. 1990a. Inheritance of resistance to anthracnose (*Colletotrichum* spp.) in pepper (*Capsicum annuum* L.) I. Genetic analysis of anthracnose resistance by diallel crosses. *Journal of the Korean Society for Horticultural Science* 31: 91–105.

Park HK, Kim BS, Lee WS. 1990b. Inheritance of resistance to anthracnose (*Colletotrichum* spp.) in pepper (*Capsicum annuum* L.). II. Genetic analysis of resistance to *Colletotrichum dematium*. *Journal of the Korean Society for Horticultural Science* 31: 207–212.

Park H-S, Lee J, Lee S-C, Yang T-J, Yoon JB. 2016. The complete chloroplast genome sequence of *Capsicum chinense* Jacq. (Solanaceae). *Mitochondrial DNA Part B* 1: 164–165.

Park M, Choi D. 2013. The structure of pepper genome. In *Genetics, Genomics and Breeding of Peppers and Eggplants* (ed. B-C Kang, C Kole). CRC Press, Taylor & Francis Group, Boca Raton, FL.

Park M, Park J, Kim S, Kwon J-K, Park HM, Bae IH, Yang T-J et al. 2012. Evolution of the large genome in *Capsicum annuum* occurred through accumulation of single-type long terminal repeat retrotransposons and their derivatives. *The Plant Journal* 69: 1018–1029.

Park S-K, Kim SH, Park HG, Yoon JB. 2009. *Capsicum* germplasm resistant to pepper anthracnose differentially interact with *Colletotrichum* isolates. *Horticulture, Environment and Biotechnology* 50: 17–23.

Paterson AH, Lander ES, Hewitt JD, Peterson S, Lincoln SE, Tanksley SD. 1988. Resolution of quantitative traits into Mendelian factors by using a complete linkage map of restriction fragment length polymorphisms. *Nature* 335: 721–726.

Perry L, Dickau R, Zarrillo S, Holst I, Pearsall DM, Piperno DR, Berman MJ et al. 2007. Starch fossils and the domestication and dispersal of chili peppers (*Capsicum* spp. L.) in the Americas. *Science* 315: 986–988.

Pickersgill B. 1971. Relationships between weedy and cultivated forms in some species of chili peppers (genus *Capsicum*). *Evolution* 25: 683–691.

Pickersgill B. 1977. Chromosomes and evolution in *Capsicum*. In *CR 3ème Congres EUCARPIA*, Vol. Capsicum 77 (ed. E Pochard), Piment, INRA, Avignon-Montfavet, France.

Pickersgill B. 1997. Genetic resources and breeding of *Capsicum* spp. *Euphytica* 96: 129–133.

Pillai S, Gopalan V, Lam AK-Y. 2017. Review of sequencing platforms and their applications in phaeochromocytoma and paragangliomas. *Critical Reviews in Oncology/Hematology* 116: 58–67.

Powell W, Machray GC, Provan J. 1996. Polymorphism revealed by simple sequence repeats. *Trends in Plant Science* 1: 215–222.

Pozzobon MT, Schifino-Wittmann MT, De Bem Bianchetti L. 2006. Chromosome numbers in wild and semidomesticated Brazilian *Capsicum* L. (*Solanaceae*) species: Do x = 12 and x = 13 represent two evolutionary lines? *Botanical Journal of the Linnean Society* 151: 259–269.

Prince JP, Cheng D, Otero CF. 2013. Molecular mapping of complex traits in *Capsicum*. In *Genetics, Genomics and Breeding of Peppers and Eggplants* (ed. B-C Kang, C Kole). CRC Press, Taylor & Francis Group, Boca Raton, FL.

Qin C, Yu C, Shen X, Fang X, Chen L, Min J, Cheng J et al. 2014. Whole-genome sequencing of cultivated and wild peppers provides insights into *Capsicum* domestication and specialization. *Proceedings of the National Academy of Sciences of the United States of America* 111: 5135–5140.

Rahaman MM, Chen D, Gillani Z, Klukas C, Chen M. 2015. Advanced phenotyping and phenotype data analysis for the study of plant growth and development. *Frontiers in Plant Science* 6: 619.

Rahman M, Paterson AH. 2009. Comparative genomics in crop plants. In *Molecular Techniques in Crop Improvement,* 2nd edition (ed. SM Jain, DS Brar), pp. 23–61. Springer Netherlands, Dordrecht.

Ranathunge NP, Mongkolporn O, Ford R, Taylor PWJ. 2012. *Colletotrichum truncatum* pathosystem on *Capsicum* spp: Infection, colonization and defence mechanisms. *Australasian Plant Pathology* 41: 463–473.

Raveendar S, Jeon Y-A, Lee J-R, Lee G-A, Lee KJ, Cho G-T, Ma K-H et al. 2015a. The complete chloroplast genome sequence of Korean landrace "Subicho" pepper (*Capsicum annuum* var. *annuum*). *Plant Breeding and Biotechnology* 3: 88–94.

Raveendar S, Na Y-W, Lee J-R, Shim D, Ma K-H, Lee S-K, Chung J-W. 2015b. The complete chloroplast genome of *Capsicum annuum* var. *glabriusculum* using Illumina sequencing. *Molecules* 20: 13080–13088.

Raweerotwiboon A, Chunwongse J. 2015. Double haploid production of peppers carrying anthracnose resistance via anther culture. *AgBio Newsletter* 7: 4–9.

Raweerotwiboon A, Meekul P, Chunwongse J. 2014a. Anther culture of peppers from inter-specific backcross (*C. annuum* x *C. baccatum*) carrying anthracnose resistance. *Khon Kaen Agriculture Journal* 42 (Supplementary 3): 802–807.

Raweerotwiboon A, Meekul P, Chunwongse J. 2014b. Induction of double haploid lines from inter-specific backcross between CA758 and $BC_2F_3$758x80C_5 (1) by anther culture. *Khon Kaen Agriculture Journal* 42 (Supplementary): 808–814.

Saikai RK, Scharf S, Faloona F, Mullis KB, Horn GT, Erlich HA, Arnheim N. 1985. Enzymatic amplification of b-globin genomic sequences and restriction site analysis for diagnosis of sickle cell anemia. *Science* 230: 1350–1354.

Saini TJ, Gupta SG, Char BR, Zehr UB, Anandalakshmi R. 2016. First report of chilli anthracnose caused by *Colletotrichum karstii* in India. *New Disease Reports* 34: 6.

Sánchez-Sevilla JF, Horvath A, Botella MA, Gaston A, Folta K, Kilian A, Denoyes B et al. 2015. Diversity arrays technology (DArT) marker platforms for diversity analysis and linkage mapping in a complex crop, the octoploid cultivated strawberry (*Fragaria* × *ananassa*). *PLoS ONE* 10: e0144960.

Sanger F, Nicklen S, Coulson AR. 1977. DNA sequencing with chain-terminating inhibitors. *Proceedings of the National Academy of Sciences of the United States of America* 74: 5463–5467.

Schoen DJ, Brown AHD. 1993. Conservation of allelic richness in wild crop relatives is aided by assessment of genetic markers. *Proceedings of the National Academy of Sciences of the United States of America* 90: 10623–10627.

Scholthof K-BG. 2007. The disease triangle: Pathogens, the environment and society. *Nature Reviews Microbiology* 5: 152–156.

Schreinemachers P, Ebert AW, Wu M-H. 2014. Costing the *ex situ* conservation of plant genetic resources at AVRDC—The World Vegetable Center. *Genetic Resources and Crop Evolution* 61: 757–773.

Seeb JE, Carvalho G, Hauser L, Naish K, Roberts S, Seeb LW. 2011. Single-nucleotide polymorphism (SNP) discovery and applications of SNP genotyping in nonmodel organisms. *Molecular Ecology Resources* 11: 1–8.

Seo H-H, Park S, Park S, Oh B-J, Back K, Han O, Kim J-I et al. 2014. Overexpression of a defensin enhances resistance to a fruit-specific anthracnose fungus in pepper. *PLoS ONE* 9: e97936.

Sharma G, Shenoy BD. 2014. *Colletotrichum fructicola* and *C. siamense* are involved in chilli anthracnose in India. *Archives of Phytopathology and Plant Protection* 47: 1179–1194.

Sharma PN, Kaur M, Sharma OP, Sharma P, Pathania A. 2005. Morphological, pathological and molecular variability in *Colletotrichum capsici*, the cause of fruit rot of chillies in the subtropical region of north-western India. *Journal of Phytopathology* 153: 232–237.

Shim D, Raveendar S, Lee J-R, Lee G-A, Ro N-Y, Jeon Y-A, Cho G-T et al. 2016. The complete chloroplast genome of *Capsicum frutescens* (Solanaceae). *Applications in Plant Sciences* 4: 1600002.

Silva JRA, Chaves TP, da Silva ARG, Barbosa LdF, Costa JFO, Ramos-Sobrinho R, Teixeira RRO et al. 2017. Molecular and morpho-cultural characterization of *Colletotrichum* spp. associated with anthracnose on *Capsicum* spp. in north-eastern Brazil. *Tropical Plant Pathology* 42: 315–319.

Silva SAM, Rodrigues R, Gonçalves LSA, Sudré CP, Bento CS, Carmo MGF, Medeiros AM. 2014. Resistance in *Capsicum* spp. to anthracnose affected by different stages of fruit development during pre- and post-harvest. *Tropical Plant Pathology* 39: 335–341.

Simko I, Piepho H-P. 2012. The area under the disease progress stairs: Calculation, advantage, and application. *Phytopathology* 102: 381–389.

Sobrino B, Brión M, Carracedo A. 2005. SNPs in forensic genetics: A review on SNP typing methodologies. *Forensic Science International* 154: 181–194.

Soh HC, Park AR, Park S, Back K, Yoon JB, Park HG, Kim YS. 2012. Comparative analysis of pathogenesis-related protein 10 (*PR10*) genes between fungal resistant and susceptible peppers. *European Journal of Plant Pathology* 132: 37–48.

Sreenivasaprasad S, Brown AE, Mills PR. 1992. DNA sequence variation and interrelationship among *Colletotrichum* species causing strawberry anthracnose. *Physiological and Molecular Plant Pathology* 41: 265–281.

Srinivasan K. 2016. Biological activities of red pepper (*Capsicum annuum*) and its pungent principle capsaicin: A review. *Critical Reviews in Food Science and Nutrition* 56: 1488–1500.

Stael S, Kmiecik P, Willems P, Van Der Kelen K, Coll NS, Teige M, Van Breusegem F. 2015. Plant innate immunity—sunny side up? *Trends in Plant Science* 20: 3–11.

Sterner RT, Shumake SA, Gaddis SE, Bourassa JB. 2005. *Capsicum* oleoresin: Development of an in-soil repellent for pocket gophers. *Pest Management Science* 61: 1202–1208.

Sucheela K. 2012. Evaluation of screening methods for anthracnose disease in chilli. *Pest Management in Horticultural Ecosystems* 18: 188–193.

Sun C, Mao SL, Zhang ZH, Palloix A, Wang LH, Zhang BX. 2015. Resistances to anthracnose (*Colletotrichum acutatum*) of *Capsicum* mature green and ripe fruit are controlled by a major dominant cluster of QTLs on chromosome P5. *Scientia Horticulturae* 181: 81–88.

Sutton BC. 1992. The genus *Glomerella* and its anamorph *Colletotrichum*. In *Colletotrichum Biology, Pathology and Control* (ed. JA Bailey, MJ Jeger), pp. 1–26. CAB International, Wallingford, UK.

Suwor P, Thummabenjapone P, Sanitchon J, Kumar S, Techawongstien S. 2015. Phenotypic and genotypic responses of chili (*Capsicum annuum* L.) progressive lines with different resistant genes against anthracnose pathogen (*Colletotrichum* spp.). *European Journal of Plant Pathology* 143: 725–736.

Syukur M, Sujiprihati S, Koswara J, Widodo. 2013. Genetic analysis for resistance to anthracnose caused by *Colletotrichum acutatum* in chili pepper (*Capsicum annuum* L.) using diallel crosses. *SABRAO Journal of Breeding and Genetics* 45: 400–408.

Tanksley SD. 1993. Mapping polygenes. *Annual Review of Genetics* 27: 205–233.

Tanksley SD, Bernatzky R, Lapitan NL, Prince JP. 1988. Conservation of gene repertoire but not gene order in pepper and tomato. *Proceedings of the National Academy of Sciences of the United States of America* 85: 6419–6423.

Tanksley SD, Grandillo S, Fulton TM, Zamir D, Eshed Y, Petiard V, Lopez J et al. 1996. Advanced backcross QTL analysis in a cross between an elite processing line of tomato and its wild relative *L. pimpinellifolium*. *Theoretical and Applied Genetics* 92: 213–224.

Tariq A, Naz F, Rauf CA, Irshad G, Abbasi NA, Khokhar NM. 2016. First report of anthracnose caused by *Colletotrichum truncatum* on bell pepper (*Capsicum annuum*) in Pakistan. *Plant Disease* 101: 631.

Tautz D, Renz M. 1984. Simple sequences are ubiquitous repetitive components of eukaryotic genomes. *Nucleic Acids Research* 12: 4127–4138.

Taylor PWJ, Ford R. 2007. Diagnostics, genetic diversity and pathogenic variation of ascochyta blight of cool season food and feed legumes. *European Journal of Plant Pathology* 119: 127–133.

TeBeest DO, Correll JC, Weidemann GJ. 1997. Specification and population biology in *Colletotrichum*. In *The Mycota V, Part B.* (ed. K Esser, PA Lemke), pp. 157–168. Springer-Verlag, Berlin, Heidelberg.

Temiyakul P. 2012. Anthracnose resistance of *Capsicum baccatum* L. "PBC80": A trispecies hybrid production and a study of differential gene expressions in mature green and ripe fruit stages. PhD thesis, p. 127. *Center for Agricultural Biotechnology*, Kasetsart University, Kamphaeng Saen Campus, Thailand.

Temiyakul P, Taylor PWJ, Mongkolporn O. 2010. Development of a double-inoculation method to assess resistance to anthracnose in trispecies *Capsicum* hybrid. *Journal of Phytopathology* 158: 561–565.

Temiyakul P, Taylor PWJ, Mongkolporn O. 2012. Differential fruit maturity plays an important role in chili anthracnose infection. *The Journal of King Mongkut's University of Technology North Bangkok* 22: 494–504.

Than PP, Jeewon R, Hyde KD, Pongsupasamit S, Mongkolporn O, Taylor PWJ. 2008. Characterization and pathogenicity of *Colletotrichum* species associated with anthracnose on chilli (*Capsicum* spp.) in Thailand. *Plant Pathology* 57: 562–572.

The Plant List. 2013. Version 1.1; www.theplantlist.org/ (Date accessed 23 September 2017).

Tropicos®. 2017. Tropicos.org. Missouri Botanical Garden; www.tropicos.org/ (Date accessed 23 September 2017).

Van Cleef EAC. 2016. Are Glandular Trichomes of *Capsicum* Important for Direct and Indirect Defence? M.Sc. Thesis, p. 39. Wageningen University, Wageningen, The Netherlands.

Van Hintum TJL, Brown AHD, Spillane C, Hodgkin T. 2000. Core collections of plant genetic resources. In *IPGRI Technical Bulletin No 3*. International Plant Genetic Resources Institute, Rome, Italy.

Van Zonneveld M, Ramirez M, Williams DE, Petz M, Meckelmann S, Avila T, Bejarano C et al. 2015. Screening genetic resources of *Capsicum* peppers in their primary center of diversity in Bolivia and Peru. *PLoS ONE* 10: e0134663.

Vasanthakumari MM, Shivanna MB. 2013. Biological control of anthracnose of chilli with rhizosphere and rhizoplane fungal isolates from grasses. *Archives of Phytopathology and Plant Protection* 46: 1641–1666.

Voorrips RE, Finkers R, Sanjaya L, Groenwold R. 2004. QTL mapping of anthracnose (*Colletotrichum* spp.) resistance in a cross between *Capsicum annuum* and *C. chinense*. *Theoretical and Applied Genetics* 109: 1275–1282.

Vos P, Hogers R, Bleeker M, Reijans M, van de Lee T, Hornes M, Frijters A et al. 1995. AFLP: A new technique for DNA fingerprinting. *Nucleic Acids Research* 23: 4407–4414.

Walsh BM, Hoot SB. 2001. Phylogenetic relationships of *Capsicum* (Solanaceae) using DNA sequences from two noncoding regions: The chloroplast atpB-rbcL spacer region and nuclear waxy introns. *International Journal of Plant Science* 162: 1409–1418.

Wang D, Bosland PW. 2006. The genes of *Capsicum*. *HortScience* 41: 1169–1187.

Wang Y, Wang Y. 2018. Trick or treat: Microbial pathogens evolved apoplastic effectors modulating plant susceptibility to infection. *Molecular Plant-Microbe Interactions* 31: 6–12.

Wang YY, Sun C-S, Wang C-C, Chien N-F. 1973. The induction of the pollen plantlets of *Triticale* and *Capsicum annuum* from anther culture. *Scientia Sinica* 16: 147–151.

Weir BS, Johnston PR, Damm U. 2012. The *Colletotrichum gloeosporioides* species complex. *Studies in Mycology* 73: 115–180.

Williams JGK, Kubelik AR, Livak KJ, Rafalski JA, Tingey SV. 1990. DNA polymorphisms amplified by arbitrary primers are useful as genetic markers. *Nucleic Acids Research* 18: 6531–6535.

Wu F, Eannetta NT, Xu Y, Durrett R, Mazourek M, Jahn MM, Tanksley SD. 2009. A COSII genetic map of the pepper genome provides a detailed picture of synteny with tomato and new insights into recent chromosome evolution in the genus *Capsicum*. *Theoretical and Applied Genetics* 118: 1279–1293.

Wu F, Tanksley SD. 2010. Chromosomal evolution in the plant family Solanaceae. *BMC Genomics* 11: 182.

Yeung MF, Tang WYM. 2015. Clinicopathological effects of pepper (oleoresin capsicum) spray. *Hong Kong Medical Journal* 21: 542–552.

Yi G, Lee JM, Lee S, Choi D, Kim B-D. 2006. Exploitation of pepper EST–SSRs and an SSR-based linkage map. *Theoretical and Applied Genetics* 114: 113–130.

Yoon JB, Do JW, Kim SH, Park HG. 2009. Inheritance of anthracnose (*Colletotrichum acutatum*) resistance in *Capsicum* using interspecific hybridization. *Korean Journal of Agricultural Science* 27: 140–144.

Yoon JB, Park HG. 2005. Trispecies bridge crosses, (*Capsicum annuum* x *C. chinense*) x *C. baccatum*, as an alternative for introgression of anthracnose resistance from *C. baccatum* into *C. annuum*. *Journal of the Korean Society for Horticultural Science* 46: 5–9.

Yoon JB, Yang DC, Do JW, Park HG. 2006. Overcoming two post-fertilization genetic barriers in interspecific hybridization between *Capsicum annuum* and *C. baccatum* for introgression of anthracnose resistance. *Breeding Science* 56: 31–38.

Yoon JB, Yang DC, Lee WP, Ahn SY, Park HG. 2004. Genetic resources resistant to anthracnose in the genus *Capsicum*. *Journal of the Korean Society for Horticultural Science* 45: 318–323.

Yumnam JS, Tyagi W, Pandey A, Meetei NT, Rai M. 2012. Evaluation of genetic diversity of chilli landraces from north eastern India based on morphology, SSR markers and the Pun1 locus. *Plant Molecular Biology Reporter* 30: 1470–1479.

Zeng ZB. 1994. Precision mapping of quantitative trait loci. *Genetics* 136: 1457–1468.

Zhang X-M, Zhang Z-H, Gu X-Z, Mao S-L, Li X-X, Chadœuf Jl, Palloix A et al. 2016. Genetic diversity of pepper (*Capsicum* spp.) germplasm resources in China reflects selection for cultivar types and spatial distribution. *Journal of Integrative Agriculture* 15: 1991–2001.

Zhao W, Wang T, Chen QQ, Chi YK, Swe TM, Qi RD. 2016. First report of *Colletotrichum scovillei* causing anthracnose fruit rot on pepper in Anhui Province, China. *Plant Disease* 100: 2168.

Zietkiewicz E, Rafalski A, Labuda D. 1994. Genome fingerprinting by simple sequence repeat (SSR)-anchored polymerase chain reaction amplification. *Genomics* 20: 176–183.

Index

About the Author

Orarat Mongkolporn is an associate professor in the department of horticulture, Kasetsart University of Thailand. Her expertise is in plant breeding for disease resistance employing molecular technology. She has been working on chili anthracnose pathogenicity and breeding for resistance since 2000. Her research findings have established core knowledge on the genetics of pathogenicity and anthracnose resistance in chili.

Born in Thailand, the land of agriculture, Orarat chose to study agricultural science due to her wish to develop agriculture in her country. She was awarded an Australian scholarship to pursue a PhD and completed her doctorate in molecular breeding at the University of Melbourne, Australia, in 1998. She has developed her professional research skills with efforts to solve the anthracnose disease problem in chili, which is the subject of this book.